DOE/CE - 0384

Energy-Efficient Electric Motor Selection Handbook

Revision 3

January 1993

DOE/CE - 0384

Energy-Efficient Electric Motor Selection Handbook

Prepared by
Gilbert A. McCoy
Todd Litman
John G. Douglass
Washington State Energy Office
Olympia, Washington

Funded by
The Bonneville Power
Administration
United States
Department of Energy

Revision 3
January 1993

Acknowledgments

The authors wish to thank the U.S. Department of Energy and the Bonneville Power Administration for funding this project. Particular thanks are due to Craig Wohlgemuth and Joe Flores of Bonneville's Industrial Technology Section for their encouragement, direction, and support. Karen Meadows, Washington State Energy Office's Industrial Program Manager, Rob Gray, Vicki Zarrell, and Kim Lyons contributed energy audit information and a technical review as did Chris Picotte of the Montana Department of Natural Resources and Conservation and Artie Dewey of the Idaho Department of Water Resources. Appreciation is also extended to Austin Bonnett of U.S. Electrical Motors, Jim Raba of the National Electrical Manufacturers Association, K.K. Lobodovsky of Pacific Gas and Electric, James Arndt of Reliance Electric, Gary Schurter of Marathon Electric, Mike Prater and Andy Chien of Teco American, Michael Clemente of the Lincoln Electric Company, Terry Thome of Grainger, Bill Green of General Electric, James McCormick of Baldor Electric, John Widstrand of Leeson Electric, Harold Musolf of Harris Electric Motors, and Dennis Brown of Seattle Electrical Works for information and/or review services provided.

Disclaimer

This report was prepared by the Washington State Energy Office as an account of work sponsored by the United States Government. Neither the United States nor the Bonneville Power Administration, the State of Washington, the Washington State Energy Office, nor any of their contractors, subcontractors, or their employees, makes any warranty, expressed or implied, or assumes any legal responsibility for the accuracy, completeness, or usefulness of any information, apparatus, product, or process disclosed within the report.

BPA Report Summary

Industrial

SUBJECT

High Efficiency Electric Motors

TITLE

ENERGY-EFFICIENT ELECTRIC MOTOR SELECTION HANDBOOK

SUMMARY

Substantial reductions in energy and operational costs can be achieved through the use of energy-efficient electric motors. A handbook was compiled to help industry identify opportunities for cost-effective application of these motors. It covers the economic and operational factors to be considered when motor purchase decisions are being made. Its audience includes plant managers, plant engineers, and others interested in energy management or preventative maintenance programs.

BPA PERSPECTIVE

Conservation is a cornerstone of the Bonneville Power Administration's (BPA) resource program. A major concern of the BPA is to promote the use of cost-effective electric conservation technologies. The industrial sector is the largest of the four electric energy consuming sectors. Within the sector electric motors are the largest users of that energy. One study estimated recently that 52.7 average megawatts of electric power, valued at $13.8 million, could be saved in the Northwest through the replacement of standard motors with high efficiency models. Of the 2 million industrial motors sold nationwide each year, energy efficient motors represent only 15 percent of the sales. That figure is probably even lower in the Northwest where electricity is cheap. Assisting industry in considering the high-efficiency motor option was the goal of this project.

BACKGROUND

The efficiency of an electric motor can only be improved through a reduction in motor losses. Improvement in the design, materials, and construction have resulted in efficiency gains of 2 to 6 percent which translates into a 25 percent reduction in losses. A small gain in efficiency can produce significant energy savings and lower operating costs over the life of the motor. Consequently, the higher purchase price of high-efficiency motors (15 to 30 percent) can, in most cases, be recovered in 2 years through cost savings in energy and operation.

POWER ADMINISTRATION

Industrial Technology

Because energy-efficient motors are a proven technology in terms of durability and reliability, their use should be considered for new installations, major modifications, replacement of failed motors or those that require rewinding, or extreme cases of oversized or underloaded motors.

OBJECTIVE

To assist the industrial sector in identifying cost-effective opportunities for application of energy-efficient motors.

APPROACH

The Handbook contains a discussion on the characteristics, economics, and benefits of standard versus high-efficiency motors in the 1 to 200 horsepower range. A motor performance database is supplied for use in identifying, evaluating, and purchasing energy-efficient motors and includes information on full and part load nominal efficiency and power factor as well as material on specific models and costs. Descriptions of how operating factors such as speed and design voltage effect performance are included. Typical operating conditions are also covered. Steps are outlined for launching a motor improvement program, which includes a worksheet to determine potential energy savings and the economic feasibility of an energy-efficient motor project.

PROJECT LEAD

Craig Wohlgemuth, P.E.
Industrial Technology Section - RMID
Bonneville Power Administration
PO. Box 3621
Portland, OR 97208
(503) 230-3044

ORDERING INFORMATION

Report Number: DOE/BPA-34623-3, February 1992

For additional copies of this report, call BPA's toll free document request line, 1-800-622-4520. You will hear a recorded message; please leave your request.

Preface

Efficient use of energy enables commercial and industrial facilities to minimize production costs, increase profits, and stay competitive. The majority of electrical energy consumed in most industrial facilities is used to run electric motors. Energy-efficient motors now available are typically from 2 to 6 percent more efficient than their standard motor counterparts. This efficiency improvement translates into substantial energy and dollar savings. For instance, a recent study of Northwest industrial sector energy conservation measures revealed a potential for 52.7 MWa of energy savings by replacing standard motors with high-efficiency motors. This savings is annually valued at $13.8 million given an electricity price of only $0.03/kWh (see Chapter 5).

The price premium for an energy-efficient motor is typically 15 to 30 percent above the cost of a standard motor. Over a typical lo-year operating life, a motor can easily consume electricity valued at over 57 times its initial purchase price. This means that when you spend $1,600 to purchase a motor, you are obligating yourself to purchase over $92,000 worth of electrical energy to operate it. A price premium of $400 is negligible compared to saving 3 percent of $92,000 or $2,760. Purchasing new or replacement energy-efficient motors makes good economic sense (see Chapter 5).

Energy-efficient motors are truly premium motors. The efficiency gains are obtained through the use of refined design, better materials, and improved construction. Many motor manufacturers offer an extended warranty for their premium-efficiency motor lines. Yet only 15 percent of motor sales nationwide are of high-efficiency units. Because of our low-cost electricity, this percentage is undoubtedly even lower in the Northwest region.

Durable and reliable energy-efficient motors can be extremely cost effective with simple paybacks on investment of less than 2 years-even in the Northwest. Energy-efficient motors should be considered in the following instances.

- For new facilities or when modifications are made to existing installations or processes
- When procuring equipment packages
- Instead of rewinding failed motors
- To replace oversized and underloaded motors
- As part of an energy management or preventative maintenance program
- When utility rebates am offered that make high-efficiency motor retrofits even more cost effective

This Energy-Efficient Electric Motor Selection Handbook (Handbook) shows you how to assess energy savings and cost effectiveness when making motor purchase decisions. The Handbook also discusses high-efficiency motor speed characteristics, performance under part-load conditions, and operation with an abnormal power supply.

Additionally, the Handbook tells you where further information is available. For example, you can obtain performance and price data for both standard and energy-efficient motors from the Electric Ideas Clearinghouse 1-800-872-3568, or 206-586-8588 outside of BPA's service area. Finally, the Handbook contains a motor test data sheet (Appendix A) and a list of Northwest motor manufacturers' representatives (Appendix B).

This page intentionally left blank.

Chapter 1
Introduction

When to Buy Energy-Efficient Motors

This Energy-Efficient Electric Motor Selection Handbook (Handbook) contains guidelines to help you identify motors that are candidates for replacement with energy-efficient electric motors. Using readily available information such as motor nameplate capacity, operating hours, and electricity price you can quickly determine the simple payback that would result from selecting and operating an energy-efficient motor.

Using energy-efficient motors can reduce your operating costs in several ways. Not only does saving energy reduce your monthly electrical bill, it can postpone or eliminate the need to expand the electrical supply system capacity within your facility. On a larger scale, installing energy conserving devices allows your electrical utility to defer building expensive new generating plants, resulting in lower costs for you, the consumer.

Energy-efficient motors are higher quality motors, with increased reliability and longer manufacturer's warantees, providing savings in reduced downtime, replacement and maintenance costs.

Saving this energy and money requires the proper selection and use of energy-efficient motors.[1] There are three general opportunities for choosing energy-efficient motors: 1) when purchasing a new motor, 2) in place of rewinding failed motors, and 3) to retrofit an operable but inefficient motor for energy conservation savings. Energy-efficient motors should be considered in the following instances:[2]

- For all new installations
- When major modifications are made to existing facilities or processes
- For all new purchases of equipment packages that contain electric motors, such as air conditioners, compressors, and filtration systems
- When purchasing spares or replacing failed motors
- Instead of rewinding old, standard-efficiency motors
- To replace grossly oversized and underloaded motors
- As part of an energy management or preventative maintenance program
- When utility conservation programs, rebates, or incentives are offered that make energy-efficient motor retrofits cost-effective

Industrial Motor Populations and Uses

In 1987, industrial sector use of electricity in the Northwest amounted to 6,062 average megawatts (MWa). This is equivalent to 38.8 percent of the region's 15,618 MWa of total electricity sales to final consumers. Five industries--food, chemicals, paper, lumber, and metals-accounted for more than 90 percent of the region's industrial use of electricity[3] A 1988 study of possible industrial sector energy conservation measures revealed a potential of approximately 345 MWa of energy savings, with changeouts of standard to energy-efficient motors accounting for 52.7 MWa or 15.2 percent of the total savings.[4] Replacing standard with energy-efficient motors saves $13.8 million annually given an electricity price of only $.03/kWh.

Figure 1
US Motor Population by End-Use Application (Thousands)

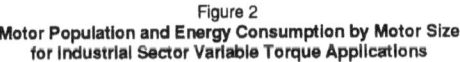

Total Applications
35,237 Units

- Variable Torque 32.5%
- Centrifugal Fans, 5,017 (14.2%)
- Centrifugal Pumps, 6,437 (18.3%)
- Positive Displacement Pumps, 12,077 (34.3%)
- Material Hdlg, 11,706 (33.2%)
- Constant Torque 67.5%

Variable Torque
11,454 Units

- Commercial Pumps, 947 (8.3%)
- Commercial Fans, 3763 (32.9%)
- Industrial Fans, 1254 (10.9%)
- Industrial Pumps, 5490 (47.9%)

Figure 2
Motor Population and Energy Consumption by Motor Size for Industrial Sector Variable Torque Applications

Energy Consumption (kWh/Yr x 10^6)
Total kWh/Yr: 170,500

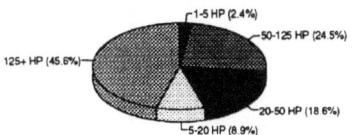

- 1-5 HP (2.4%)
- 50-125 HP (24.5%)
- 125+ HP (45.6%)
- 20-50 HP (18.6%)
- 5-20 HP (8.9%)

Population (Thousands)
Total Units: 6,744

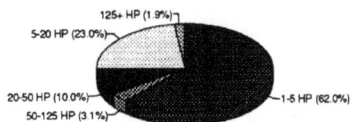

- 125+ HP (1.9%)
- 5-20 HP (23.0%)
- 20-50 HP (10.0%)
- 50-125 HP (3.1%)
- 1-5 HP (62.0%)

A nationwide survey by a major motor manufacturer indicates that approximately 35.2 million motors larger than 1 horsepower (hp) are used within the commercial and industrial sectors.[5] Approximately 6.7 million units drive centrifugal pumps and fans in industrial applications while over 23 million motors are used for constant torque conveyor and positive displacement pumping applications. Figure 1 breaks down motor population by end-use.

While 62 percent of the motors used in variable torque centrifugal pump and fan applications in the industrial sector are in the 1- to 5-hp size range, these small motors account for only 2.4 percent of total motor energy consumption. Motors in the 5- to 125-hp size range use 52 percent of the sector's energy, while large motors account for 45.6 percent of electricity consumption. Figure 2 summarizes motors and energy consumption by motor size.[5]

Motors are the largest single use of electricity in most industrial plants. A study conducted for Seattle City Light indicates that 42 percent of that utility's industrial customer electrical consumption goes to motor driven end uses.[6] The dominance of motor loads can be even greater in some industries. For instance, energy audits conducted by the Washington State Energy Office reveal the following motor loads at various manufacturing plants. A saw and planer mill has 65 motors with 6,215 hp of connected load. Approximately 94 percent of the 13.8 million kWh of that facility's annual electrical energy consumption goes to motors driving boiler feedwater pumps, forced draft fans, hydraulic systems, air compressors, planer drives, blowers, feeders, chippers, edgers, hoggers, debarkers, radial saws, and slabbers. A small cedar mill similarly has 37 motors with 2,672 total hp. A Northwest plywood drying facility uses 72 motors with 3,275 hp of nameplate capacity. These motors drive combustion air fans, scrubbers, circulating air fans, condensate pumps, charging pumps, hoggers, fines and chip blowers, bag house blowers, and glue mixers. Forty-seven percent of the electrical consumption at a controlled-atmosphere cold storage facility is due to refrigeration system compressor, evaporator fan, and condenser fan motors while a potato processing plant has 17 motors with 1,115 hp driving ammonia compressors, hydraulic pumps, and air compressors.

Annual Electric Motor Sales Volume

A major manufacturer estimates that U.S. annual sales exceed 2 million motors. Table 1 describes sales volume by motor horsepower. Only 15 percent of these sales involve premium or high-efficiency motors.[7] A survey of motor manufacturers conducted by the Washington State Energy Office (WSEO) indicates that the highest priorities of motor buyers are availability/quick delivery, reliability, and price. Energy efficiency was ranked 5th out of 11 purchaser concerns.

Table 1
Polyphase Induction Motor Annual Sales Volume

hp	Units (Thousands)
1-5	1,300
7-1/2 - 20	500
25-50	140
60-100	40
125-200	32
250-500	11
Total	2,023

It is likely that the percentage of energy-efficient motor sales is even lower in the Pacific Northwest. Motor dealers state that..."customers don't ask for efficiency data and we don't tell them...", "...I haven't sold a high-efficiency motor in a year and a half...," "...high-efficiency motors just don't make sense with the low electrical rates we have in the Northwest...," and "...customers get turned off when we tell them that high-efficiency motors have a price premium of 15 to 20 percent and an efficiency improvement of only 2 to 3 percent"

Contrary to "common knowledge," energy-efficient motor purchases can be quite cost effective with simple paybacks on investment of less than 2 years-even in the Northwest.

Chapter 2
Energy-Efficient Motor Performance and Price

The efficiency of a motor is the ratio of the mechanical power output to the electrical power input. This may be expressed as:

$$Efficiency = \frac{Output}{Input} = \frac{Input-Losses}{Input} = \frac{Output}{Output + Losses}$$

Design changes, better materials, and manufacturing improvements reduce motor losses, making premium or energy-efficient motors more efficient than standard motors. Reduced losses mean that an energy-efficient motor produces a given amount of work with less energy input than a standard motor.[2]

In 1989, the National Electrical Manufacturers Association (NEMA) developed a standard definition for energy-efficient motors. The definition, designed to help users identify and compare electric motor efficiencies on an equal basis, includes a table of minimum nominal full-load efficiency values.* A motor's performance must equal or exceed the nominal efficiency levels given in Table 2 for it to be classified as "energy-efficient"

Nominal full-load efficiencies for currently available energy-efficient and standard motors are shown in Figure 3. Figure 3 clearly indicates that the NEMA standards are easy for motor manufacturers to exceed. In fact, most motors on the market qualify as "high-efficiency" machines. It is also apparent that you can improve efficiency by as much as 6 points through simply buying a "premium-efficiency" motor, one whose performance lies near the top of the range of available efficiencies, rather than one that just meets the NEMA minimum standard.

Frequently, one manufacturer's energy-efficient motor performs with approximately the same efficiency as another manufacturer's standard unit. Average nominal efficiencies and 1990 list prices for standard and energy-efficient motors are summarized in Table 3.

In order to help you identify, evaluate, and procure energy-efficient motors, the Washington State Energy Office has prepared a motor performance database. The database contains full- and part-load nominal efficiencies and power factors for approximately 2,700, 5- to 300-hp NEMA Design B polyphase motors. Information contained in the database was extracted from manufacturers' catalogs with each manufacturer given an opportunity to review performance information for accuracy. Users can query the database to produce a listing, ranked in order of descending full-load efficiency, for all motors within a stated size, speed, and enclosure classification.

A sample database listing is shown in Table 4. The database also contains the manufacturers' name, motor model, full-load RPM, service factor, frame size, and list price. Note that the nominal full-load motor efficiencies vary from 86.5 to 93.2 percent. Prices also vary. In many cases, motors with identical list prices exhibit very different efficiency ratings.

Table 2
NEMA Threshold Nominal Full Load Efficiencies for Energy Efficient Motors

12-6B

HP	ODP				TEFC			
	3600	1800	1200	900	3600	1800	1200	900
1	82.5	77	72	80.5	75.5	72		
1.5	80	82.5	82.5	75.5	78.5	81.5	82.5	75.5
2	82.5	82.5	84	85.5	82.5	82.5	82.5	82.5
3	82.5	86.5	85.5	86.5	82.5	84	84	81.5
5	85.5	86.5	86.5	87.5	85.5	85.5	85.5	84
7.5	85.5	88.5	88.5	88.5	85.5	87.5	87.5	85.5
10	87.5	88.5	90.2	89.5	87.5	87.5	87.5	87.5
15	89.5	90.2	89.5	89.5	87.5	88.5	89.5	88.5
20	90.2	91	90.2	90.2	88.5	90.2	89.5	89.5
25	91	91.7	91	90.2	89.5	91	90.2	89.5
30	91	91.7	91.7	91	89.5	91	91	90.2
40	91.7	92.4	91.7	90.2	90.2	91.7	91.7	90.2
50	91.7	92.4	91.7	91.7	90.2	92.4	91.7	91
60	93	93	92.4	92.4	91.7	93	91.7	91.7
75	93	93.6	93	93.6	92.4	93	93	93
100	93	93.6	93.6	93.6	93	93.6	93	93
125	93	93.6	93.6	93.6	93	93.6	93	93.6
150	93.6	94.1	93.6	93.6	93	94.1	94.1	93.6
200	93.6	94.1	94.1	93.6	94.1	94.5	94.1	94.1

12-6C

HP	ODP				TEFC			
	3600	1800	1200	900	3600	1800	1200	900
1	82.5	80	74	75.5	82.5	80	74	
1.5	82.5	84	84	75.5	82.5	84	85.5	77
2	84	84	85.5	85.5	84	84	86.5	82.5
3	84	86.5	86.5	86.5	85.5	87.5	87.5	84
5	85.5	87.5	87.5	87.5	87.5	87.5	87.5	85.5
7.5	87.5	88.5	88.5	88.5	88.5	89.5	89.5	85.5
10	88.5	89.5	90.2	89.5	89.5	89.5	89.5	88.5
15	89.5	91	90.2	89.5	90.2	91	90.2	88.5
20	90.2	91	91	90.2	90.2	91	90.2	89.5
25	91	91.7	91.7	90.2	91	92.4	91.7	89.5
30	91	92.4	92.4	91	91	92.4	91.7	91
40	91.7	93	93	91	91.7	93	93	91
50	92.4	93	93	91.7	92.4	93	93	91.7
60	93	93.6	93.6	92.4	93	93.6	93.6	91.7
75	93	94.1	93.6	93.6	93	94.1	93.6	93
100	93	94.1	94.1	93.6	93.6	94.5	94.1	93
125	93.6	94.5	94.1	93.6	94.5	94.5	94.1	93.6
150	93.6	95	94.5	93.6	94.5	95	95	93.6
200	94.5	95	94.5	93.6	95	95	95	94.1

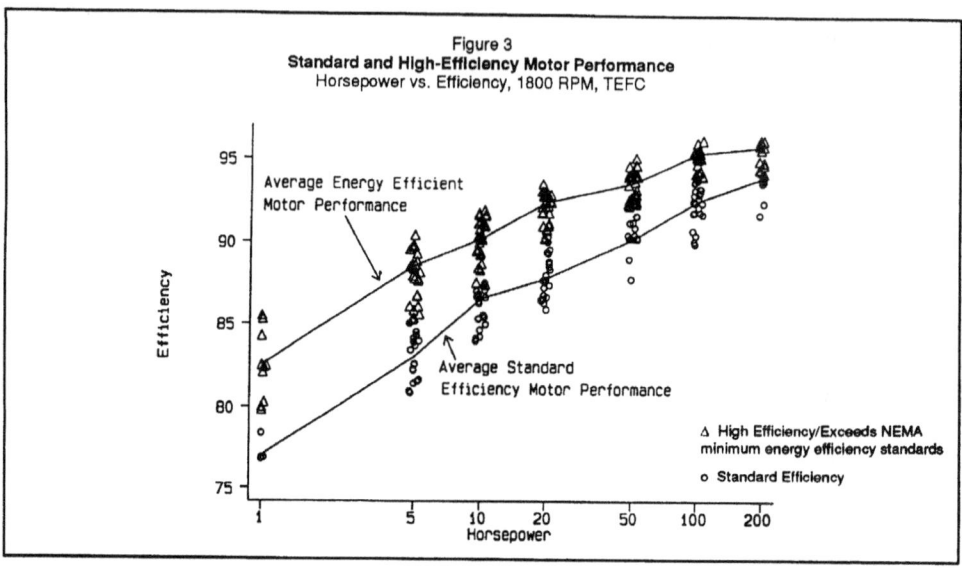

Figure 3
Standard and High-Efficiency Motor Performance
Horsepower vs. Efficiency, 1800 RPM, TEFC

Motor Losses and Loss Reduction Techniques

A motor's function is to convert electrical energy to mechanical energy to perform useful work. The only way to improve motor efficiency is to reduce motor losses. Even though standard motors operate efficiently, with typical efficiencies ranging between 83 and 92 percent, energy-efficient motors perform significantly better. An efficiency gain from only 92 to 94 percent results in a 25 percent reduction in losses. Since motor losses result in heat rejected into the atmosphere, reducing losses can significantly reduce cooling loads on an industrial facility's air conditioning system.

Motor energy losses can be segregated into five major areas, each of which is influenced by design and construction decisions.[9] One design consideration, for example, is the size of the air gap between the rotor and the stator. Large air gaps tend to maximize efficiency at the expense of power factor, while small air gaps slightly compromise efficiency while significantly improving power factor.[10] Motor losses may be categorized as those which are fixed, occurring whenever the motor is energized, and remaining constant for a given voltage and speed, and those which are variable and increase with motor load.[11] These losses are described below.

1. **Core loss** represents energy required to magnetize the core material (hysteresis) and includes losses due to creation of eddy currents that flow in the core. Core losses are decreased through the use of improved permeability electromagnetic (silicon) steel and by lengthening the core to reduce magnetic flux densities. Eddy current losses are decreased by using thinner steel laminations.

2. **Windage and friction** losses occur due to bearing friction and air resistance. Improved bearing selection, air-flow, and fan design are employed to reduce these losses. In an energy-efficient motor, loss minimization results in reduced cooling requirements so a smaller fan can be used. Both core losses and windage and friction losses are independent of motor load.

Table 3
Average Efficiencies and Typical List Prices
for Standard and Energy-Efficient Motors
1800 RPM Open Drip-Proof Motors

hp	Average Standard Motor Efficiency, %	Average Energy-Efficient Motor Efficiency, %	Efficiency Improvement, %	Typical Standard ODP Motor List Price	Typical Energy-Efficient ODP Motor List Price	List Price Premium
5	83.8 (15)	87.9 (12)	4.7	$329 (4)	$370 (4)	$41
7.5	85.3 (14)	89.6 (15)	4.8	408 (6)	538 (5)	130
10	87.2 (21)	91.1 (7)	4.3	516 (6)	650 (5)	134
15	87.6 (15)	91.5 (11)	4.3	677 (5)	864 (5)	187
20	88.4 (14)	92.0 (11)	3.9	843 (6)	1055 (5)	212
25	89.2 (14)	92.8 (11)	3.9	993 (5)	1226 (5)	233
30	89.2 (12)	92.8 (12)	3.9	1160 (4)	1425 (5)	265
40	90.2 (12)	93.6 (11)	3.6	1446 (4)	1772 (5)	326
50	90.1 (11)	93.6 (13)	3.7	1688 (6)	2066 (4)	378
60	91.0 (11)	94.1 (12)	3.3	2125 (7)	2532 (5)	407
75	91.9 (11)	94.5 (12)	2.8	2703 (5)	3084 (5)	381
100	91.7 (9)	94.5 (14)	3.0	3483 (6)	3933 (5)	450
125	91.7 (7)	94.4 (16)	2.9	4006 (6)	4709 (5)	703
150	92.9 (8)	95.0 (12)	2.2	5760 (5)	6801 (5)	1041
200	93.1 (8)	95.2 (12)	2.2	7022 (3)	8592 (3)	1570

1800 RPM Totally Enclosed Fan-Cooled Motors

hp	Average Standard Motor Efficiency, %	Average Energy-Efficient Motor Efficiency, %	Efficiency Improvement, %	Typical Standard TEFC Motor List Price	Typical Energy-Efficient TEFC Motor List Price	List Price Premium
5	83.3 (11)	87.3 (32)	4.6	$344 (6)	$448 (5)	$104
7.5	85.2 (20)	89.5 (22)	4.8	494 (7)	647 (5)	153
10	86.0 (10)	89.4 (30)	3.8	614 (6)	780 (5)	166
15	86.3 (8)	90.4 (27)	4.5	811 (7)	1042 (5)	231
20	88.3 (13)	92.0 (20)	4.0	1025 (6)	1268 (5)	243
25	89.3 (14)	92.5 (19)	3.5	1230 (7)	1542 (5)	312
30	89.5 (9)	92.6 (23)	3.3	1494 (6)	1824 (5)	330
40	90.3 (10)	93.1 (21)	3.0	1932 (7)	2340 (5)	408
50	91.0 (9)	93.4 (22)	2.6	2487 (5)	2881 (4)	394
60	91.7 (11)	94.0 (19)	2.4	3734 (7)	4284 (5)	514
75	91.6 (6)	94.1 (24)	2.7	4773 (7)	5520 (5)	747
100	92.1 (13)	94.7 (17)	2.7	5756 (5)	6775 (4)	1019
125	92.0 (10)	94.7 (19)	2.9	7425 (5)	9531 (5)	2106
150	93.0 (10)	95.0 (18)	2.1	9031 (6)	11123 (3)	2092
200	93.8 (9)	95.4 (14)	1.7	10927 (5)	13369 (4)	2442

Note: Full-load efficiencies are given. The numbers in parenthesis indicate either the number of motors considered or number of motor manufacturers using the identical list price. List prices are extracted from 1990 manufacturers' brochures.

Table 4: Typical MotorMaster Database Report

11/25/92 MotorMaster Database Query

CRITERIA Horsepower. 20
          Speed (RPM). 1800
          Enclosure. . . . . Totally Enclosed
          Voltage . . . . 230 V

Manufacter	Model	F.L. Eff.	F.L. P.F.	F.L. RPM	Frame	Catalog #	List Price
US Motors	PREMIUM EFFICIENCY	93.3	84.9	1770	256T	A443	1268
Reliance	XE PREMIUM EFF.	93.1	84.5	1768	256T	P25G3331	1345
Baldor	SUPER-E	93.0	85.0	1760	256T	EM2334T	1266
Toshiba	E.Q.P.	93.0	84.2	1772	256T	80204FLF2UMH	1268
Teco	MAX-E1/HE	93.0	83.0	1765	256T	1535	
Sterling	SILVERLINE	93.0	88.0	1760	256T	JH0204FFA	1263
Marathon	BLUE CHIP XRI	93.0	83.0	1770	256T	E206	1268
Magnetek	E-PLUS III	93.0	86.5	1765	256T	E464	1268
GE	ENERGYSAVER	93.0	82.5	1765	256T	E963	1268
GE	EXTRA $EVERE DUTY	93.0	82.5	1765	256T	E9154	1535
Sterling	J SERIES, U-FRAME	92.4	84.3	1775	286U	JU0204FFA	1844
Siemens	PE-21 SEVERE DUTY	92.4	89.2	1760	256T	HP13780	1535
Sterling	U SERIES/ HIGH EFF	92.4	84.3	1775	286U	0	
Baldor	WASHDOWN-C FACE	91.7	82.0	1760	256TC	CWDM23934T	1880
Magnetek	E-PLUS	91.7	88.0	n/a	S256T	E421	1559
Leeson	C-FLEX, CAST IRON	91.2	n/a	1750	256TC	150089	1125
Sterling	C-FACE, FOOT MOUNT	91.0	85.0	1745	256TC	JB0204FHA	1237
Sterling	K SERIES	91.0	85.0	1760	256T	KB0204FFA	895
Magnetek	STD EFF	91.0	85.0	1750	256T	T448	1025
NEMA	** 12-6C STANDARD	91.0	0.0	0	———	SEPT. 1990	0
Reliance	E-2000 ENERGY EFF.	90.5	81.2	1757	256T	P25G3151	1233
US Motors	HOSTILE DUTY	90.2	86.1	1745	256T	E409	1025
GE	STD EFF	90.2	84.5	1760	256T	N308	1121
NEMA	** 12-6B STANDARD	90.2	0.0	0	———	NOV. 1989	0
US Motors	CLOSE COUPLED	89.8	85.4	1765	256	B050	942
Toshiba	STD EFF	89.6	89.0	1751	256T	B0204FLF2UD	1025
Delco	C.I.M/ENERGY EFF.	89.6	75.5	1755	256T	1535	
US Motors	HIGH EFFICIENCY	89.5	84.9	1760	256T	A939	1025
Marathon	BLUE CHIP	89.5	84.5	1750	256T	H438	1025
Dayton	CAST IRON HAZ. LOC	89.5	84.5	1750	256T	3N491	1465
Siemens	STD EFF	89.0	88.0	1750	256T	HT13710	1025
Delco	T LINE	88.6	78.9	1750	256T		1025
Baldor	TEFC-C FACE	88.5	81.0	1760	256TC	VM2334T	1137
Baldor	TEFC-RIGID BASE	88.5	85.0	1760	286U	M2334	1911
Teco	STD EFF	88.5	90.5	1750	256T		1025
Sterling	U SERIES	88.5	89.4	1760	286U	1612	
Reliance	STD EFF	87.9	85.1	1753	256T	P25G312	1025
Magnetek	IRON HORSE	87.5	85.6	1750	256T	N480	1493
Lincoln	LINCOLN	86.5	86.6	1750	256T	KM180.230460	718

Purchase Price Versus Running Cost - A Comparison

Let's compare the fuel cost savings of an efficient automobile over a less efficient automobile with savings obtained from purchase of an energy-efficient over a standard-efficiency motor. Based upon 15,000 miles per year at a fuel economy of 25 miles per gallon with gasoline priced at $1.00 per gallon, the fuel cost of a typical car is $600 per year or about 6.6 percent of the $9,000 purchase price. A 5-mile-per-gallon improvement in fuel economy saves 100 gallons of gasoline valued at $100 annually.

In contrast, a 15-hp standard efficiency motor, continuously operating at 75 percent of its full rated load, would consume 85,189 kWh/year of electrical energy. At an electricity rate of only $.03/kWh, this energy is valued at $2,555 or 315 percent of the motor's list price. A typical 15-hp continuously operating energy-efficient motor conserves 3,863 kWh of electricity valued at $116 annually. Vehicle and motor purchase alternatives are summarized below.

Vehicle Versus Motor Purchase—Comparison Base Case

New Car (25 MPG)		New 15-hp Standard-Efficiency Motor	
Purchase Price:	$9,000	List Price:	$811
Drive:	15,000 miles/yr	Use:	Continuous
MPG:	25	Load Factor:	75 Percent
Gal/Yr:	600	kWh/Yr:	85,189
Fuel Cost Value @ $1.00/gal:	$600	Electricity Cost @ .03/kWh:	$2,555
Ratio of Annual Fuel Cost To Initial Cost:	6.6 Percent		315 Percent

Alternatives

Fuel-Efficient Car (30 MPG)		New 15-hp Energy-Efficient Motor	
Efficiency Improvement:	5 MPG	Efficiency Improvement: 86.3 to 90.4 percent	
Annual Fuel Savings:	100 Gal.	Annual kWh Savings:	3,863
Value of Savings:	$100/year	Value of Savings:	$116/year
Savings over 150,000 mile operating life: 1,000 gallons, $1,000		Savings over 20-year service life: 77,260 kWh and $2,320	

Over a 20-year operating period, the standard motor would consume approximately 1.7 million kWh of electrical energy. This energy is valued at $51,100 or more than 6,300 percent of the initial motor purchase price.

3. **Stator losses** appear as heating due to current flow (I) through the resistance (R) of the stator winding. This is commonly referred to as an I^2R loss. I^2R losses can be decreased by modifying the stator slot design or by decreasing insulation thickness to increase the volume of wire in the stator.

4. **Rotor losses** appear as I^2R heating in the rotor winding. Rotor losses can be reduced by increasing the size of the conductive bars and end rings to produce a lower resistance, or by reducing the electrical current.

5. **Stray load losses** are the result of leakage fluxes induced by load currents. Both stray load losses and stator and rotor I^2R losses increase with motor load. Motor loss components are summarized in Table 5. Loss distributions as a function of motor horsepower are given in Table 6 while variations in losses due to motor loading are shown in Figure 4.[13,14]

Table 5
Motor Loss Categories

No Load Losses	Typical Losses (%)	Factors Affecting these Losses
Core Losses	15 -25	Type and quantity of magnetic material
Friction and Windage Losses	5 - 15	Selection and design of fans and bearings
Motor Operating Under Load		
Stator I^2R Losses	25 - 40	Stator conductor size
Rotor I^2R Losses	15 - 25	Rotor conductor size
Stray Load Losses	10 - 20	Manufacturing and design methods

Table 6
**Typical Distributions of Motor Losses, %
(1800 RPM Open Drip-Proof Enclosure)**

Types of Loss	Motor Horsepower		
	25	50	100
Stator I^2R	42	38	28
Rotor I^2R	21	22	18
Core Losses	15	20	13
Windage and Friction	7	8	14
Stray Load	15	12	27

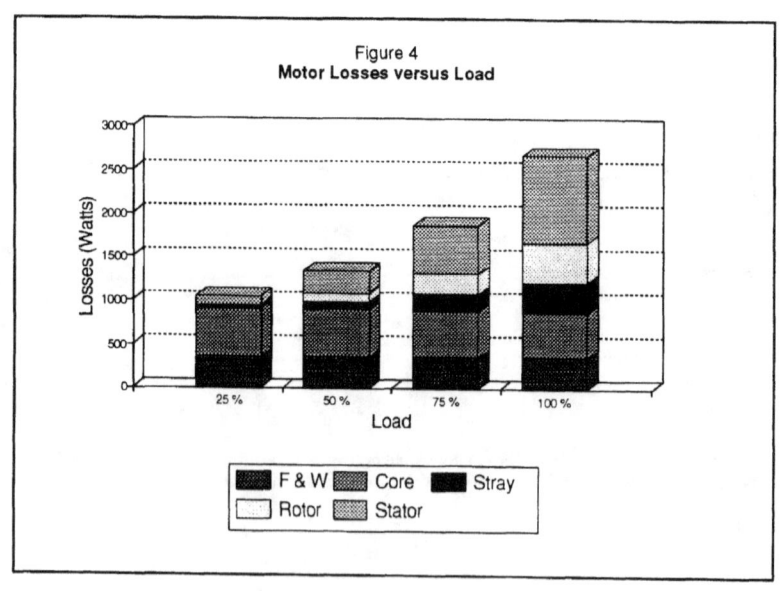

Figure 4
Motor Losses versus Load

Determining and Comparing Motor Efficiencies

Efficiency Definitions Vary

When evaluating motors on the basis of efficiency improvements or energy savings, it is essential that a uniform efficiency definition be used. It is often difficult to accurately compare manufacturers' published, quoted, or tested efficiencies, as various values are used in catalogues and vendor literature. Common definitions include[11]

- **Average or Nominal Efficiency.** These terms are identical and refer to the average full-load efficiency value obtained through testing a sample population of the same motor model. These are the most common standards used to compare motors.
- **Guaranteed Minimum or Expected Minimum Efficiency.** All motors purchased or a stated percentage of the motors purchased are guaranteed to have efficiencies that equal or exceed this full-load value. (Based on NEMA table.)
- **Apparent Efficiency.** "Apparent efficiency" is the product of motor power factor and minimum efficiency. With this definition energy consumption can vary considerably as the power factor can be high while the efficiency is low. Specifications should not be based on "apparent" efficiency values.
- **Calculated Efficiency.** This term refers to an average expected efficiency based upon a relationship between design parameters and test results. Specifications should not be based on "calculated" efficiency values.

Motor Efficiency Testing Standards

It is critical that motor efficiency comparisons be made using a uniform product testing methodology. There is no single standard efficiency testing method that is used throughout the industry.[2,10] The most common standards are:

- IEEE 112 -1984 (United States)
- IEC 34-2 (International Electrotechnical Commission)
- JEC - 37 (Japanese Electrotechnical Committee)
- BS - 269 (British)
- C-390 (Canadian Standards Association)
- ANSI C50.20 same as IEEE 112 (United States)

IEEE Standard 112- 1984, *Standard Test Procedure for Polyphase Induction Motors and Generators,* is the common method for testing induction motors in the United States. Five methods for determining motor efficiency are recognized. The common practice for motors in the l- to 125-hp size range is to measure the motor power output directly with a dynamometer while the motor is operating under load. Motor efficiency is then determined by carefully measuring the electrical input and the mechanical power output.[10]

The motor efficiency testing standards differ primarily in their treatment of stray load losses. The Canadian Standards Association (CSA) methodology and IEEE 112 - Test Method B determine the stray load loss through an indirect process. The IEC standard assumes stray load losses to be fixed at 0.5 percent of input, while the JEC standard assumes there are no stray load losses.[2] As indicated in Table 7, the efficiency of a motor, when tested under the different standard conventions, can vary by several percentage points.[2,15]

Table 7
Efficiency Results From Various Motor Testing Standards

Standard	Full-Load Efficiency (%)	
	7.5 hp	20 hp
Canadian (CSA C390)	80.3	86.9
United States (IEEE-112, Test Method B)	80.3	86.9
International (EC - 34.2)	82.3	89.4
British (BS - 269)	82.3	89.4
Japanese (JEC - 37)	85.0	90.4

Testing Equipment Accuracy Limitations

Each motor manufacturer has its own test equipment. The accuracy of this equipment varies. Three types of dynamometers are commonly used for testing medium- and large-scale motors: eddy current clutches, water brakes, and direct current generators. These units have different speed and accuracy limitations.[10] Instrumentation in use during the motor test can also affect accuracy.

The variation in testing ability was illustrated by a round robin test sponsored by the National Electrical Manufacturers Association (NEMA) in 1978. Three motors of different sizes (5, 25, and 100 hp) were shipped to nine different motor manufacturers with the request that they be tested in accordance with IEEE 112 - Method B practices.[6] A second test examined efficiency changes due to variations in materials and manufacturing tolerances. This exercise involved testing motors of a common design that were manufactured over a period of months. These tests show that variation in measured losses frequently exceed ± 10 percent for specific motor designs while the combined variation from manufacturing and testing with state-of-the-art techniques can exceed ± 19 percent. Test results are given in Table 8.[9]

What does an efficiency uncertainty mean to motor purchasers? It means that a motor rated at 92.5 percent efficiency is essentially comparable to one with a 92.0 percent value. Purchasers of energy-efficient motors should select a unit that has a measured efficiency value within 1.5 percent or less of the maximum value available within a given size, enclosure, and speed class. Motors that barely exceed the NEMA minimum qualifying standards are not recommended.

Table 8
Uncertainty in Full-Load Efficiency Measurements due to Manufacturing Variations

Motor hp	IEEE 112-Method B Testing of Identical Motors
5	87.1 ± 0.7 percent
25	89.5 ± 0.8 percent
100	91.9 ± 0.9 percent

NEMA Motor Nameplate Labeling Standards

NEMA instituted a nameplate labeling standard (MG1-12.542) for NEMA Design A and B polyphase induction motors in the 1 to 125 hp size range. The stamped full-load motor nameplate efficiency is selected from a table of nominal efficiencies and represents a value that is "not greater than the average efficiency of a large population of motors of the same design," tested in accordance with IEEE 112. In addition, the full-load efficiency, when the motor is operated at rated voltage and frequency, shall not be less than the minimum value associated with a nominal value in Table 9.
Efficiencies shall be identified on the nameplate by the caption "NEMA Nominal Efficiency" or "NEMA Nom. Eff."[10]

NEMA nominal efficiency bands are selected to "avoid the inference of undue accuracy that might be assumed from using an infinite number of nominal efficiency values."[9] The efficiency bands vary between 0.4 to 1.5 percent between the 84 and 95.5 percent nominal efficiency range. Motors with efficiencies falling within a given band may be treated as having essentially equivalent operating efficiencies. The nameplate nominal efficiency thus represents a value that may be used to compare the relative energy consumption of a motor or group of motors."

Table 9
NEMA Motor Nameplate Efficiency Marking Standard

Nominal Efficiency (%)	Minimum Efficiency (%)	Nominal Efficiency (%)	Minimum Efficiency (%)
98.0	97.6	87.5	85.5
97.8	97.4	86.5	84.0
97.6	97.1	85.5	82.5
97.4	96.8	84.0	81.5
97.1	96.5	82.5	80.0
96.8	96.2	81.5	78.5
96.5	95.8	80.0	77.0
96.2	95.4	78.5	75.5
95.8	95.0	77.0	74.0
95.4	94.5	75.5	72.0
95.0	94.1	74.0	70.0
94.5	93.6	72.0	68.0
94.1	93.0	70.0	66.0
93.6	92.4	68.0	64.0
93.0	91.7	66.0	62.0
92.4	91.0	64.0	59.5
91.7	90.2	62.0	57.5
91.0	89.5	59.5	55.0
90.2	88.5	57.5	52.5
89.5	87.5	55.0	50.5
88.5	86.5	52.5	48.0

Chapter 3
How Much Can You Save?

The amount of money you can save by purchasing an energy-efficient motor instead of a standard motor depends on motor size, annual hours of use, load factor, efficiency improvement, and the serving utility's charges for electrical demand and energy consumed

Three pieces of information are required to evaluate the economic feasibility of procuring an energy-efficient motor instead of a standard motor. First, obtain a copy of your utility's rate schedule. Then determine load factor or percentage of full rated output. Finally, determine the number of motor operating hours at rated load With this information you can determine your annual energy and cost savings.

Understanding Your Utility's Rate Schedule

The cost of electricity for a commercial or industrial facility is typically composed of four components:

1. **Basic or Hookup Charge.** A fixed amount per billing period that is independent of the quantity of electricity used This charge covers the cost of reading the meter and servicing your account.

2. **Energy Charges.** A fixed rate ($/kWh) or rates, times the electrical consumption (kWh) for the billing period. Energy charges are frequently seasonally differentiated and may also vary with respect to the quantity of electricity consumed. Utility tariffs may feature declining block or inverted rate schedules. With a declining block rate schedule, illustrated in Table 10, energy unit prices decrease as consumption increases.

3. **Demand Charge.** A fixed rate ($/kW) times the billable demand (kW) for the billing period. Demand charges are often based upon the highest power draw for any 15-minute time increment within the billing period. Some utilities feature ratcheted demand charges where the applicable monthly demand charge is the highest value incurred during the preceding year.

4. **Power Factor Penalty or Reactive Power Charge.** A penalty is frequently levied if power factor falls below an established value (typically 90 or 95 percent). A low power factor indicates that a facility is consuming a proportionally larger share of reactive power. While reactive power (VAR) does not produce work and is stored and discharged in the inductive and capacitive elements of the circuit, distribution system or I^2R losses occur. The utility requires compensation for these losses.

Table 10
Utility Rate Schedule Showing
Seasonal Pricing and Declining Block Rates

Monthly Rate:	
Basic Charge:	$4.55 for single-phase or $19.00 for three-phase service.
Demand Charge:	No charge for the first 50 kW of billing demand. $5.35 per kW for all over 50 kW of billing demand.

Energy Charge:	
October - March	April - September
5.2156	4.9672 cents per kWh for the first 20,000 kWh
4.1820	3.9829 cents per kWh for the next 155,000 kWh
2.9695	2.8281 cents per kWh for all over 175,000 kWh

Determining Load Factor

Secondly, determine the load factor or average percentage of full-rated output for your motor. To calculate the load factor, compare the power draw (obtained through watt meter or voltage, amperage, and power factor measurements) with the nameplate rating of the motor. For a three-phase system, wattage draw equals the product of power factor and volts times amps times 1.732.

Determining Operating Hours

Lastly, determine the number of motor operating hours at rated load. Electrical energy savings are directly proportional to the number of hours a motor is in use. All things being equal, a high-efficiency motor operated 8,000 hours per year will conserve four times the quantity of energy of an equivalent motor that is used 2,000 hours per year.

Determining Annual Energy Savings

Before you can determine the annual dollar savings, you need to estimate the annual energy savings.

Energy-efficient motors require fewer input kilowatts to provide the same output as a standard-efficiency motor. The difference in efficiency between the high-efficiency motor and a comparable standard motor determines the demand or kilowatt savings. For two similar motors operating at the same load, but having different efficiencies, the following equation is used to calculate the kW reduction.[2,10]

Equation 1

$$kW_{saved} = hp \times L \times 0.746 \times \left(\frac{100}{E_{std}} - \frac{100}{E_{HE}}\right)$$

where:
- hp = Motor nameplate rating
- L = Load factor or percentage of full operating load
- E_{std} = Standard motor efficiency under actual load conditions
- E_{HE} = Energy-efficient motor efficiency under actual load conditions

The kW savings are the demand savings. The annual energy savings are calculated as follows:[2]

Equation 2

$$kWh_{savings} = kW_{saved} \times \text{Annual Operating Hours}$$

You can now use the demand savings and annual energy savings with utility rate schedule information to estimate your annual reduction in operating costs. Be sure to apply the appropriate seasonal and declining block energy charges.

The total annual cost savings is equal to:

Equation 3

Total savings =
$(kW_{saved} \times 12 \times \text{monthly demand charge}) +$
$(kWh_{savings} \times \text{energy charge})$

The above equations apply to motors operating at a specified constant load. For varying loads, you can apply the energy savings equation to each portion of the cycle where the load is relatively constant for an appreciable period of time. The total energy savings is then the sum of the savings for each load period. Determine the demand savings at the peak load point. The equations are not applicable to motors operating with pulsating loads or for loads that cycle at rapidly repeating intervals!'

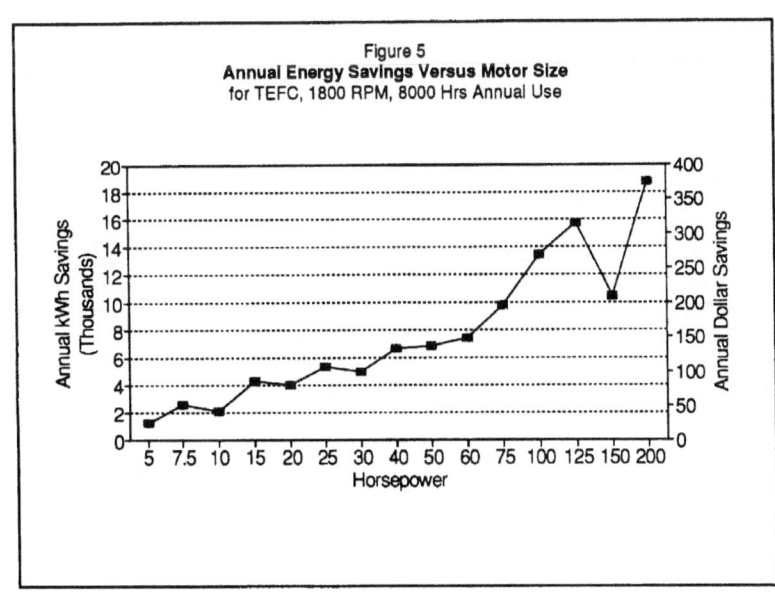

Figure 5
Annual Energy Savings Versus Motor Size
for TEFC, 1800 RPM, 8000 Hrs Annual Use

Savings also depend on motor size and the gain in efficiency between a new high-efficiency motor and a new or existing standard efficiency unit. Energy-efficient motor savings, based upon an average energy charge of $0.04/kWh, are shown in Figure 5. The performance gain for the energy-efficient motor is based on the difference between the average nominal full-load efficiencies for all energy-efficient motors on the market as compared to the average efficiency for all standard-efficiency units.

Motor Purchase Prices

Motor dealers rarely sell motors at the manufacturer's full list price. Even a customer walking in "off the street" would be offered a list price discount. Motor prices continuously vary, and rather than reprint catalogs and brochures, manufacturers advertise high list prices and authorize their dealers to provide discounts. Several major manufacturers tend to use the same list prices, given in Table 3, for both their lines of standard and energy-efficient motors. Each motor manufacturer, however, has a unique discounting policy, which typically varies with respect to dealer sales volume.

The discounting practice of one motor manufacturer is given in Table 11. The dealer's wholesale price is the list price times the appropriate multiplier for the dealer's sales volume. The dealer makes it's profit through "marking up" the manufacturer's discounted list price. Typical dealer markups range from 10 to 25 percent and depend on dealership practices and the size of the purchase order or number of motors a customer buys. There is no difference in the discount for energy-efficient and standard motors. Thus, you can buy a standard or energy-efficient motor for 55 to 85 percent of the manufacturers stated list price. Be sure to get quotes from vendors and use discounted motor prices or price premiums when determining the cost effectiveness of energy-efficient motor investments.

Table 11
Typical Motor Wholesale Pricing Practices

Annual Dealer Sales Volume	List Price Multiplier (%)
0 - $35,000/year	70
$35,001-100,000/year	57
$100,000/year or more	50

Assessing Economic Feasibility

Because of better design and use of higher quality materials, energy-efficient motors cost 15 to 30 percent more than their standard efficiency counterparts. In many cases, however, this price premium is quickly recovered through energy cost savings. To determine the economic feasibility of installing energy-efficient motors, assess the total annual energy savings in relation to the price premium.

Common methods of assessing the economic feasibility of investment alternatives include:

- Simple payback
- Life cycle costing methodologies
  - Net Present Value (NPV)
  - Benefit to Cost Ratio
  - Internal Rate of Return (ROR)

Most industrial plant managers require that investments be recovered through energy savings within 1 to 3 years based on a simple payback analysis. The simple payback is defined as the period of time required for the savings from an investment to equal the initial or incremental cost of the investment. For initial motor purchases or the replacement of burned-out and unrewindable motors, the simple payback period for the extra investment associated with an energy-efficient motor purchase is the ratio of the price premium less any available utility rebate, to the value of the total annual electrical savings.

Equation 4

$$\text{Simple payback years} = \frac{\text{Price premium - utility rebate}}{\text{Total annual cost savings}}$$

For replacements of operational motors, the simple payback is the ratio of the full cost of purchasing and installing a new energy-efficient motor relative to the total annual electrical savings.

Equation 5

$$\text{Simple payback years} = \frac{\text{New motor cost + installation charge - utility rebate}}{\text{Total annual cost savings}}$$

Example:
The following analysis for a 75 hp TEFC motor operating at 75 percent of full rated load illustrates how to determine the cost effectiveness of obtaining an energy-efficient versus a standard-efficiency motor for the initial purchase case.

Kilowatts saved:

$$kW_{saved} = hp \times Load \times 0.746 \times \left(\frac{100}{E_{std}} - \frac{100}{E_{HE}}\right)$$

$$= 75 \times .75 \times 0.746 \times \left(\frac{100}{91.6} - \frac{100}{94.1}\right)$$

$$= 1.21$$

Where E_{std} and E_{HE} are the efficiencies of the standard motor and the alternative energy-efficient unit.

This is the amount of energy conserved by the energy-efficient motor during each hour of use. Annual energy savings are obtained by multiplying by the number of operating hours at the indicated load.

Energy saved:

$kWh_{savings}$ = Hours of operation x kW_{saved}
= 8,000 hours x 1.21
= 9,680 kWh/year

Annual cost savings:
Total cost savings =
(kW_{saved} x 12 x Monthly demand charge) +
($kWh_{savings}$ x Energy charge)
= 1.21 x 12 x $5.35/ kw + 9,680 x $0.03/ kWh
= $368

In this example, installing an energy-efficient motor reduces your utility billing by $368 per year. The simple payback for the incremental cost associated with a energy-efficient motor purchase is the ratio of the discounted list price premium (from Table 3) or incremental cost to the total annual cost savings. A list price discount of 75 percent is used in this analysis.

Cost Effectiveness

$$Simple\ pay\ back = \frac{List\ Price\ premium \times Discount\ factor}{Total\ annual\ cost\ savings}$$

$$= \frac{\$747 \times 0.75}{\$368} = 1.5\ years$$

Thus, the additional investment required to buy this energy-efficient motor would be recovered within 1.5 years. Energy-efficient motors can rapidly "pay for themselves" through reduced energy consumption. After this initial payback period, the annual savings will continue to be reflected in lower operating costs and will add to your firm's profits.[2]

Recommendations for Motor Purchasers

As a motor purchaser you should be familiar with and use consistent sets of nomenclature. You should also refer to standard testing procedures. Be sure to:[10]

- Insist that all guaranteed quotations are made on the same basis (i.e., nominal or guaranteed minimum efficiency).
- Prepare specifications that identify the test standard to be used to determine motor performance.
- Recognize the variance in manufacturing and testing accuracy and establish a tolerance range for acceptable performance.
- Comparison shop.
- Obtain an energy-efficient motor with a nominal efficiency within 1.5 percent of the maximum value available within an enclosure, speed, and size class.

Energy consumption and dollar savings estimates should be based upon a comparison of nominal efficiencies as determined by IEEE 112 - Method B for motors operating under appropriate loading conditions. Note that the NEMA marking standard only refers to efficiency values stamped on the motor nameplate. In contrast, manufacturers' catalogues contain values derived from dynamometer test data. When available, use catalog information to determine annual energy and dollar savings.

Making the Right Choice

Comparison shop when purchasing a motor, just as you would when buying other goods and services. Other things being equal, seek to maximize efficiency while minimizing the purchase price. Frequently, substantial efficiency gains can be obtained without paying a higher price. Figure 6 illustrates the list price versus full-load efficiency for currently marketed 10 hp/ 1800 RPM standard and energy-efficient motors. It is readily

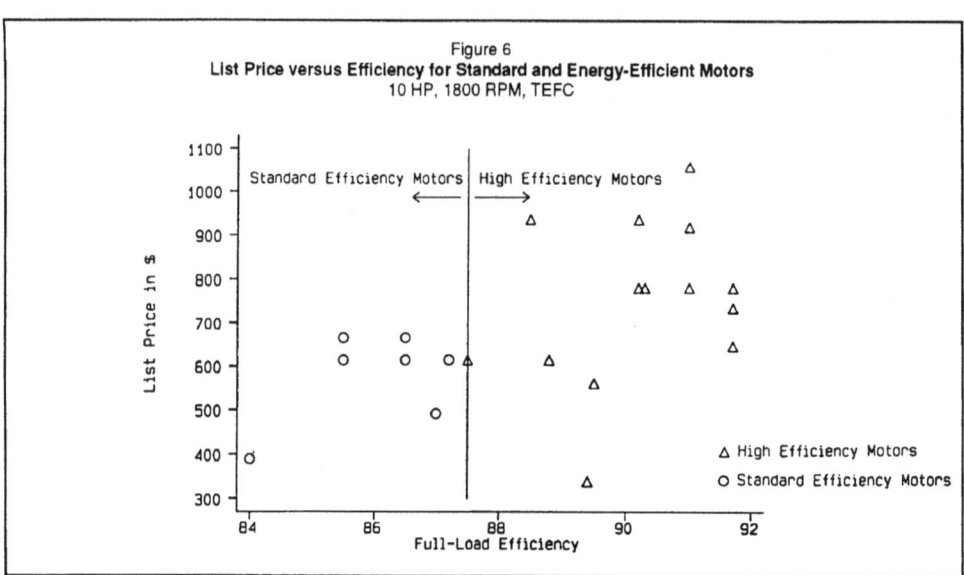

Figure 6
List Price versus Efficiency for Standard and Energy-Efficient Motors
10 HP, 1800 RPM, TEFC

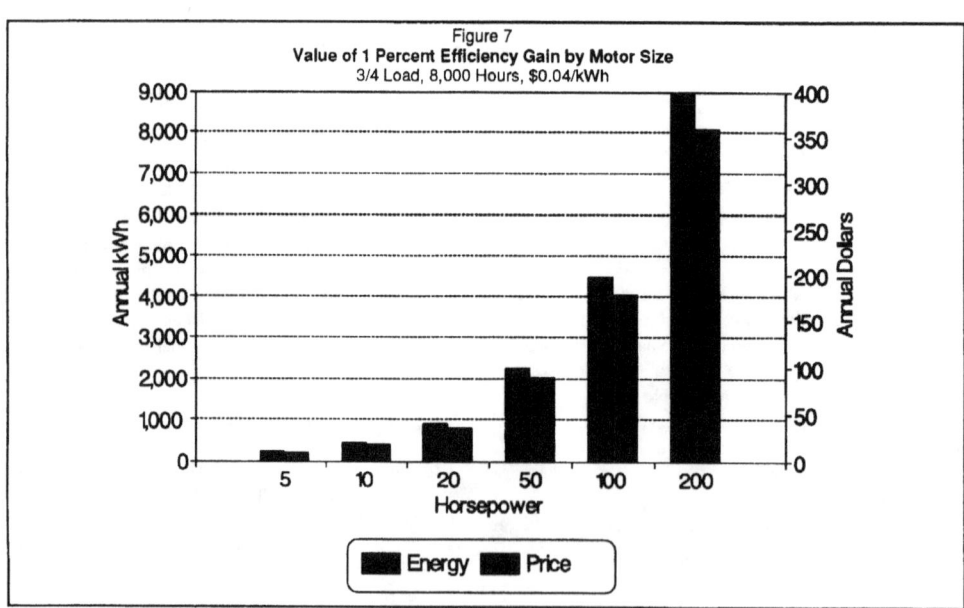

Figure 7
Value of 1 Percent Efficiency Gain by Motor Size
3/4 Load, 8,000 Hours, $0.04/kWh

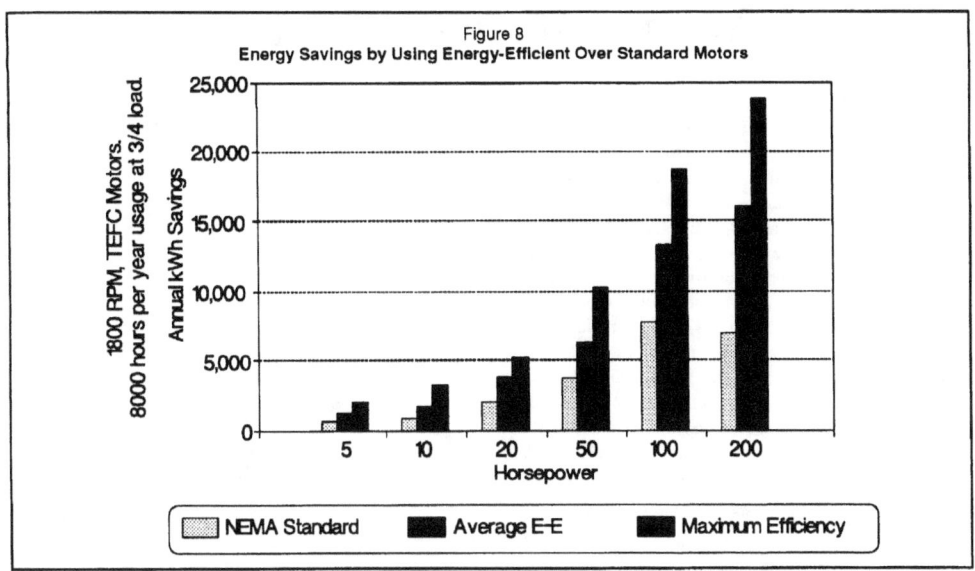

Figure 8
Energy Savings by Using Energy-Efficient Over Standard Motors

Note: Figure 8 illustrates the annual energy savings available through selection of an energy-efficient TEFC motor that just satisfies the NEMA minimum energy-efficient motor standards; for a motor that exhibits average high-efficiency performance; and for a motor with superior performance for a given speed, enclosure, and size class. The base case is the purchase of a "typical" standard-efficiency motor. Base case and average energy-efficient motor efficiencies are taken from Table 3.

apparent that you can obtain an efficiency improvement of as much as 6 points without paying any price penalty.

With the right information, you can quickly identify a motor that produces substantial energy and cost savings for little or no extra investment. The value of a 1-point efficiency improvement is shown with respect to motor horsepower in Figure 7. At an electricity price of $.04/kWh, a single point of efficiency gain for a 50 hp motor can result in an annual savings of approximately 2,600 kWh, worth $104.

Because so many motors exceed the minimum NEMA energy-efficiency standards, it is not enough to simply specify a "high-efficiency" motor. Be certain to purchase a true "premium-efficiency" motor, an energy-efficient motor with the highest available efficiency characteristics.

The value associated with "making the right choice" is graphically characterized by the minimum/maximum savings analysis illustrated in Figure 8. You can often **triple** the available savings by choosing a motor with the top performance in its class instead of a motor that barely satisfies NEMA minimum energy-efficiency standards.

Chapter 4
Obtaining Motor Efficiency Data

When choosing between rewinding a failed motor versus purchasing a new energy-efficient motor or replacing an operable standard-efficiency motor with an energy-efficient model, you can obtain annual energy savings and cost-effectiveness data by assuming that the existing motor operates at the average efficiency for all standard-efficiency units currently on the market (see Table 3) or by interpolating from Table 12. Table 12 indicates changes in full-load motor efficiency over time.[11] It is evident that efficiencies of the smaller sized mid- 1960s era T-frame motors decreased relative to the U-Frame machines available in the 1950s. The efficiency of premium efficiency motors improved dramatically in the 1980s as energy costs increased and a demand for energy-conserving motors developed.

Table 12
History of Motor Efficiency Improvements

hp	1944 Design (%)	1955 U Frame (%)	1965 Nominal Efficiency (%)	1981 Premium Efficiency (%)	1991 GE Premium Efficiency (%)
7.5	84.5	87.0	84.0	91.0	91.7
15	87.0	89.5	88.0	92.4	93.0
25	89.5	90.5	89.0	93.6	94.1
50	90.5	91.0	91.5	94.1	94.5
75	91.0	90.5	91.5	95.0	95.4
100	91.5	92.0	92.0	95.0	96.2

By using a multimeter and contact or stroboscopic tachometer you can measure the voltage, amperage, RPM, power factor, and power draw for a motor under its normal operating conditions. The slip, or difference between the synchronous and the operating speed for the motor, can be used to estimate the output kW, process load, and subsequently, the efficiency of the motor. An Induction Motor Test Data Sheet is included as Appendix A.

The synchronous speed of an induction motor depends on the frequency of the power supply and on the number of poles for which the motor is wound. The higher the frequency, the faster a motor runs. The more poles the motor has, the slower it runs.[16] The synchronous speed (N_s) for a squirrel-cage induction motor is given by Equation 6. Typical synchronous speeds are indicated in Table 13.

Equation 6
$$N_s = \frac{60 \times 2f}{p}$$

where:
f = frequency of the power supply
p = poles for which the motor is wound

Table 13
Induction Motor Synchronous Speeds

Poles	Motor Synchronous Speed, RPM 60 Hertz	50 Hertz
2	3,600	3,000
4	1,800	1,500
6	1,200	1,000
8	900	750
10	720	600
12	600	500

The actual speed of the motor is less than its synchronous speed. This difference between the synchronous and actual speed is referred to as slip. Slip is typically expressed as a percentage where?

$$Percent\ slip = \frac{(Synchronous\ speed - Actual\ speed) \times 100}{Synchronous\ speed}$$

You can now estimate motor load and efficiency with slip measurements.

Equation 7
$$Slip = RPM_{sync} - RPM_{measured}$$
$$Motor\ load = \frac{Slip}{RPM_{sync} - RPM_{full\ load\ (nameplate)}}$$

Equation 8

Approximate Output hp = Motor Load x Nameplate hp

Motor efficiency = $\dfrac{(0.746 \times Output\ hp)}{Measured\ input\ kW}$

An example:[17]

Given: RPM_{sync} = 1,800 $RPM_{measured}$ = 1,770

$RPM_{nameplate}$ = 1,750   Nameplate hp = 25

Measured kW = 13.1

Then: Slip = 1,800 - 1,770 = 30

Motor load = $\dfrac{30}{1800 - 1750} = \dfrac{30}{50}$ = 0.6

Output hp = 0.6 x 25 = 15

Motor efficiency = $\dfrac{(0.746 \times 15) \times 100\ percent}{13.1}$ = 85 percent

Slip versus load curves can be obtained from your motor manufacturer. The slip technique for determining motor load and operating efficiency should not be used with rewound motors or with motors that are not operated at their design voltage.

It should be emphasized that the slip technique is limited in its accuracy. While it cannot provide "exhaustive field efficiency testing results" it can be employed as a useful technique to identify and screen those motors that are oversized, underloaded, and operating at less than desired efficiencies.

Chapter 5
Energy-Efficient Motor Selection Guidelines

Initial Motor Purchases

When considering a new motor purchase, you have two choices: a less expensive standard-efficiency motor or a more costly energy-efficient motor. For this case, the cost premium of buying an energy-efficient motor over a standard motor is equal to the difference in prices. Installation costs are the same for both motors.

Consider a 50 hp standard-efficiency motor, purchased for $1,620 and operated 8,000 hours per year at 75 percent of full rated load. At an efficiency of 91.1 percent and with an electrical rate of $0.03 per kilowatt hour, the motor will consume $7,369 worth of electricity each year. During a typical lo-year motor operating life with an average 5 percent escalation in energy prices, the total electrical bill for operating this motor would exceed $92,600, over 50 times the purchase price of the motor.

While the improvement in efficiency associated with the purchase of an energy-efficient motor is typically only 2 to 5 percent, the incremental cost of the energy-efficient motor can often be rapidly recovered. This occurs because the price premium may be smaller than anticipated (in some cases you can actually purchase an energy-efficient motor for a lower price than the comparable standard-efficiency unit) and the ratio of the motor's annual operating cost to its initial purchase price is quite high.

Although the energy and dollar savings associated with buying an energy-efficient motor can be impressive, selecting the energy-efficient unit is not always appropriate. Motors that are lightly loaded or infrequently used, such as motors driving control valves or door openers, may not consume enough electricity to make the energy-efficient alternative cost-effective. Remember that for a motor operating under a constant load, the electricity savings associated with an efficiency improvement are directly proportional to the hours of operation.

The simple payback is defined as the period of time required for the profit or savings from an investment decision to equal the incremental cost of the investment. This investment repayment period is directly dependent upon electricity costs-for example, the simple payback for a motor operating in a utility territory where electrical rates are $.02/kWh will be twice that for a similar motor used where electricity is priced at $.04/kWh.

A short payback period indicates that an investment is worthwhile. Different industries or facilities use different simple payback criteria or investment "hurdle rates." Industrial sector energy conservation measures typically must have simple paybacks in the 1- to 3-year range to be considered cost effective.

To help you decide whether or not to purchase a new energy-efficient motor, Figure 9 presents the minimum annual operating hours required to achieve 2- and 3-year simple paybacks as a function of motor size. Simple paybacks of 2 and 3 years are considered given an electrical rate of $.04/kWh. This analysis uses the average standard- and energy-efficient motor performance data presented in Table 3 and assumes a 75 percent motor loading with a list price discount factor of 75 percent.

Several conclusions can be drawn from this analysis. In the 5- to 100-hp size range and without the availability of a utility rebate, an energy-efficient motor should be selected given the following:

- An average electricity cost of $.04/kWh; a 3-year simple payback criteria; and operation under load for more than 3,500 hours per year
- $.02/kWh electrical rate coupled with a 3-year simple payback criteria or a $.03/kWh rate with a 2-year payback with at least 7,000 hours per year of operation

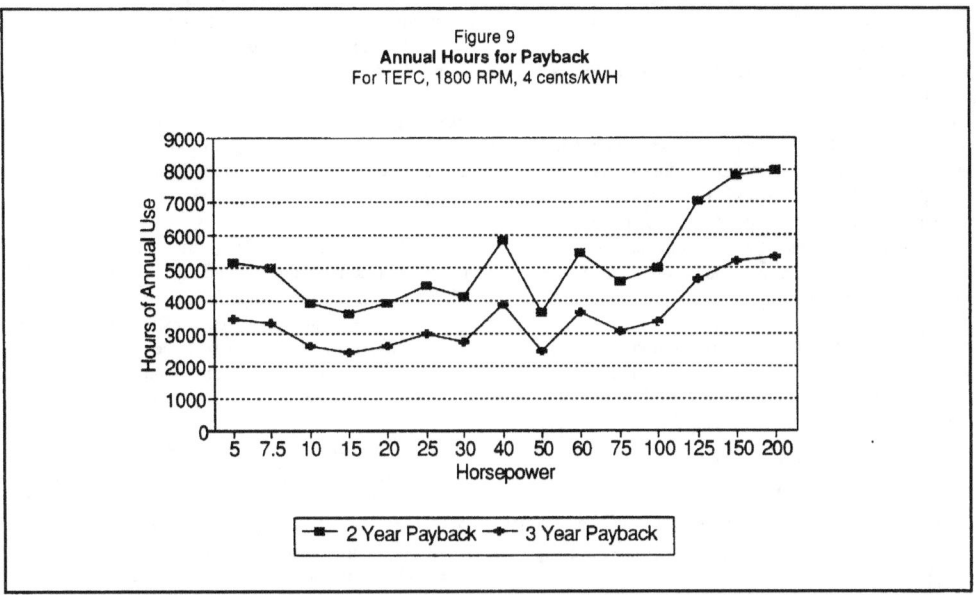

Figure 9
Annual Hours for Payback
For TEFC, 1800 RPM, 4 cents/kWH

Motor Failure and Rewind Scenario

Unlike an initial motor purchase where your decision is limited to procuring a standard versus a premium efficiency motor, a motor failure or burnout produces three alternatives. Your options are to rewind the failed motor, purchase a new standard-efficiency motor, or purchase an energy-efficient replacement motor. For this scenario, motor installation labor costs are again not included as the existing motor must be removed and reinstalled anyway.

Assuming that the failed motor can be rewound, the baseline or lowest initial cost approach is to rewind the motor to its original specifications. As some older U-Frame motors were built with oversized slots, it is sometimes possible to perform a "high-efficiency" rewind and slightly increase the efficiency of the motor by adding more copper to reduce I^2R losses.[11] If the original unit was wound with aluminum wire, it should be replaced with copper.[18]

A motor should be rewound with the same (or larger) winding wire size and configuration. If a repair shop does not have the correct wire size in stock and uses a smaller diameter wire, stator I^2R losses will increase.

While a decrease in the number of turns in a stator winding reduces the winding resistance, it also shifts the point at which the motor's peak efficiency occurs toward higher loads and increases the motor's magnetic field, starting current, locked rotor, and maximum torque. A change from 10 to 9 turns will increase the starting current by 23 percent, which can cause problems in the electrical distribution and motor protection systems.[19]

In a typical rewind, the stator is heated to a temperature high enough to burn out its winding insulation. The windings are then removed and replaced.[20] In the past, many rewind shops emphasized speed High temperatures were used to shorten repair times and get the motor back in service quickly. Hand-held torches were sometimes used to soften varnish for easier coil removal.[11,18] The resulting higher temperatures increase losses by changing the electrical characteristics of the motor's core.

For both standard and high-efficiency motors, the rewind shop should follow the motor manufacturers' recommended burnout temperature specifications. When stripping out the old windings, it is essential to keep the stator core below 700°F. If the stator core gets too hot, the insulation between the stator laminations will break down, increasing eddy current losses and lowering the motor's operating efficiency. After being damaged, the lamination insulation cannot be repaired nor the efficiency loss restored without under going a major repair such as restacking the iron. The motor also becomes less reliable.[18,20]

Insulation removal techniques vary between rewind shops and should be investigated prior to deciding where to have the motor rewound. Always choose a shop with a controlled temperature winding burnout oven to minimize core loss. Some shops have core loss testers and can screen motors determine if they are repairable prior to stripping.[18]

The repair shop should also determine and eliminate the cause for a motor's failure. Aside from proper stripping procedures, the motor owner should ensure that the rewind shop does the following."

- Uses proper methods of cleaning
- Installs Class F or better insulation
- Uses phase insulation between all phase junctions
- Uses tie and blocking methods to ensure mechanical stability
- Brazes rather than crimps connections
- Uses proper lead wire and connection lugs
- Applies a proper varnish treatment

As motor design characteristics (such as slot geometry and configuration), failure modes, rewind practices, and materials specifications and treatments vary, it is impossible to identify a "typical" rewind cost for a motor with a given horsepower, speed, and enclosure.

Motor efficiency losses after rewinds also vary considerably. While dynamometer tests conducted by the Electrical Apparatus Service Association indicate that new motors, when properly stripped and rewound, can be restored to their original efficiency, field tests on motors from a variety of manufacturing plants indicate that losses are typically higher in motors that have been rewound-perhaps because of thermal shock suffered during the motor failure.

An analysis of core loss tests taken over a 1-year period in General Electric repair facilities indicates that average core losses are 32 percent higher than normal for motors that had been previously rewound." General Electric also conducted a test of 27 rewound motors in the 3- to 150-hp size range. The test indicates that total losses increased by 18 percent for motors that have been rewound compared to those that have not been rewound.[20] An 18 percent increase in losses corresponds to an approximate 1.5 to 2.5 percent decrease in full-load efficiency.

Rewound motors can exhibit severe efficiency losses, especially if they were rewound more than 15 years ago or have been rewound several times. Rewind losses of 5 percent or more are possible.

When should a energy-efficient motor be purchased in lieu of rewinding a failed standard-efficiency motor? This decision is quite complicated as it depends on such variables as the rewind cost, expected rewind loss, energy-efficient motor purchase price, motor horsepower and efficiency, load factor, annual operating hours, electricity price, and simple payback criteria.

At least some of the time, rewinding will be the best decision. The prospects for a good rewind are greatly improved if you keep good records on your motors and provide them to the repair shop. Repair shops often can't get complete specifications from manufacturers. They must "reverse engineer" motors, counting winding turns, noting slot patterns, measuring wire size, etc. before removing old windings. Sometimes a motor has failed repeatedly in the past because of a previous non-standard rewind. The same error can be repeated unless the shop knows the motor is a "repeat offender" and diagnoses the problem. Similarly, a motor is sometimes subjected to unusual service requirements, e.g., frequent starts, dirty environment, low voltage. Most shops know how to modify original specifications to adjust to such conditions.

Here are several rewind "rules of thumb":
- Always use a qualified rewind shop. A quality rewind can maintain original motor efficiency. However, if a motor core has been damaged or the rewind shop is careless, significant losses can occur.
- Motors less than 100 hp in size and more than 15 years old (especially previously rewound motors) often have efficiencies significantly lower then current models. It is usually best to replace them.
- If the rewind cost exceeds 65 percent of a new energy-efficient motor price, buy the new motor. Increased reliability and efficiency should quickly recover the price premium.
- If your energy costs average $0.03/kWh or more, and a motor is operated for at least 4,000 hour per year, an energy-efficient motor is a worthwhile

investment. The higher purchase price will usually be repaid through energy savings within 2 years. Here is a chart to help decide when to select an energy efficient motor:

Choose a new energy-efficient motor if:

Your energy costs are:	and annual hours of use equals or exceeds:
$0.02/kWh	6,000
$0.03/kWh	4,000
$0.04/kWh	3,000
$0.05/kWh	2,000

Table 14 indicates how breakeven rewind costs vary with respect to motor operating hours and simple payback criteria. The breakeven cost is expressed as a percentage of a replacement energy-efficient motor price. A new energy-efficient motor should be purchased if the rewind cost exceeds the stated breakeven point. Table 14 may be used for NEMA Design B motors in the 5- to 125-hp size range. Assumptions used in the preparation of this table include an expected 2 percent loss in an average standard motor efficiency due to rewinding, replacement with an average energy-efficient motor operated at a 75 percent load factor, and a list price discount rate of 65 percent.

Table 14
Breakeven Rewind Cost as a Percentage of an Energy-Efficient Motor Price

Simple Payback Criteria, Years[1]	Annual Operating Hours		
	8,000	6,000	3,000
3	30%	45%	65%
2	5%	65%	70%

[1]For an electrical rate of $.03/kWh.

You can easily complete a cost-effectiveness analysis for a rewinding. If you can be assured that the past and prospective rewinds comply with all the foregoing recommended practices, the original efficiency could be maintained. Otherwise, two points should be subtracted from your standard motor efficiency to reflect expected rewind losses. Annual energy and cost savings are determined by inputting the appropriate energy-efficient motor performance, operating hours, electricity price, and load factor into Equations 1 through 3. The incremental cost of procuring the premium-efficiency unit is the quoted price for the new motor less the rewind price and any utility rebate. The simple payback for the energy-efficient motor is simply the incremental cost divided by the total annual energy conservation benefits.

Replacement of Operable Standard-Efficiency Motors

This motor retrofit scenario occurs when you replace an existing, operable standard-efficiency motor with an energy-efficient unit to conserve energy. In this instance, the cost of replacement is the full purchase price for the new motor minus any utility rebate and the salvage value for the motor to be replaced. An assumed installation cost of $500 is also levied. No downtime or loss of production costs are incurred as it is assumed that you can schedule the retrofit during a periodic maintenance shutdown. For this scenario, the entire cost of purchasing and installing the energy-efficient motor must be returned through the energy savings achieved by the increased motor efficiency.

Based on average standard and premium-efficiency motor efficiencies, simple paybacks were determined for 5-, 20-, and 100-hp motors, operating 8,000 hours per year with electricity prices of $.02 and $.03/kWh. As indicated in Figure 10, simple paybacks typically exceed 10 years. Paybacks are substantially longer for the smaller 5-hp motor due to the assumption of a fixed installation cost. Based solely on energy savings, industrial users would typically find it not cost-effective to retrofit operable standard-efficiency motors with energy-efficient units. Such an action may, however, make sense if:

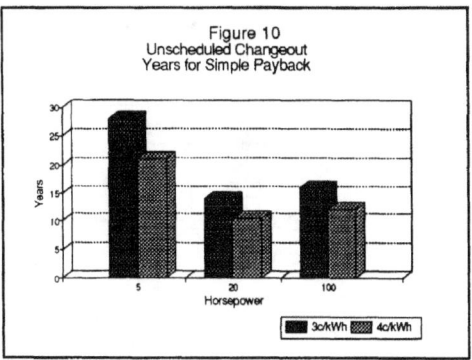

Figure 10
Unscheduled Changeout
Years for Simple Payback

- Funding is available through a utility energy conservation program to partially offset the purchase price of the new energy-efficient motor.
- The standard-efficiency motor has been rewound several times.
- The standard-efficiency motor is oversized and underloaded.

Oversized and Underloaded Motors

When a motor has a significantly higher rating than the load it is driving, the motor operates at partial load. When this occurs, the efficiency of the motor is reduced. Motors are often selected that are grossly underloaded and oversized for a particular job.

For instance, field measurements made at four industrial plants in Northern California indicate that, on the average, motors are operating at 60 percent of their rated load. The energy conservation recommendation for 17 out of the 29 motors tested was downsizing or replacement with a smaller energy-efficient motor.[21] Despite the fact that oversized motors reduce energy efficiency and increase operating costs, industries use oversized motors:[1,2]

- To ensure against motor failure in critical processes
- When plant personnel do not know the actual load and thus select a larger motor than necessary
- To build in capability to accommodate future increases in production
- To conservatively ensure that the unit has ample power to handle load fluctuations
- When maintenance staff replace a failed motor with the next larger unit if one of the correct size is not available
- When an oversized motor has been selected for equipment loads that have not materialized
- When process requirements have been reduced
- To operate under adverse conditions such as voltage imbalance

As a general rule, motors that are undersized and overloaded have a reduced life expectancy with a greater probability of unanticipated downtime, resulting in loss of production. On the other hand, motors that are oversized and thus lightly loaded suffer both efficiency and power factor reduction penalties.

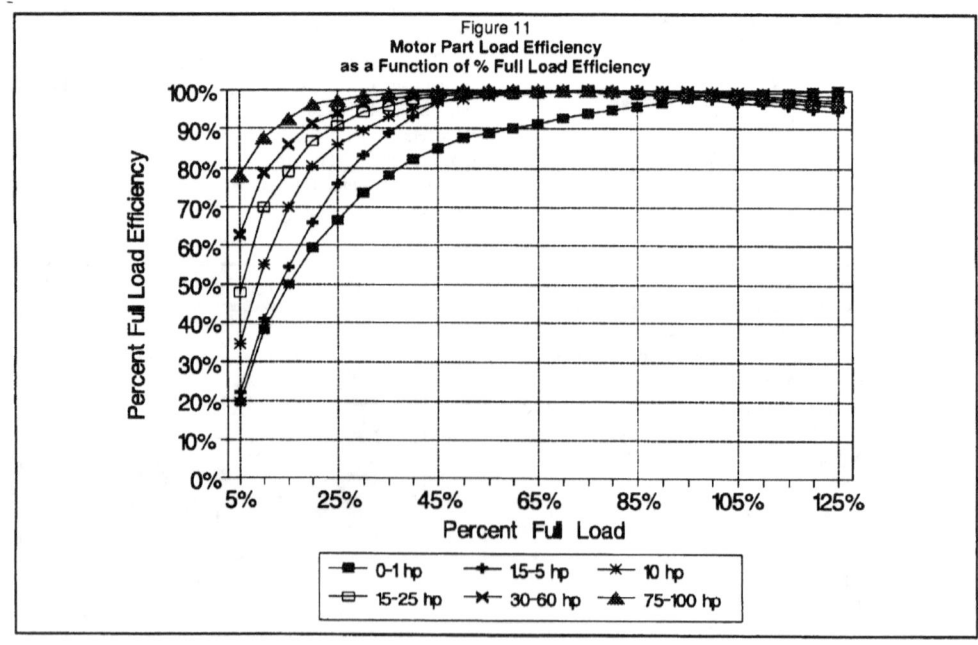

Figure 11
Motor Part Load Efficiency as a Function of % Full Load Efficiency

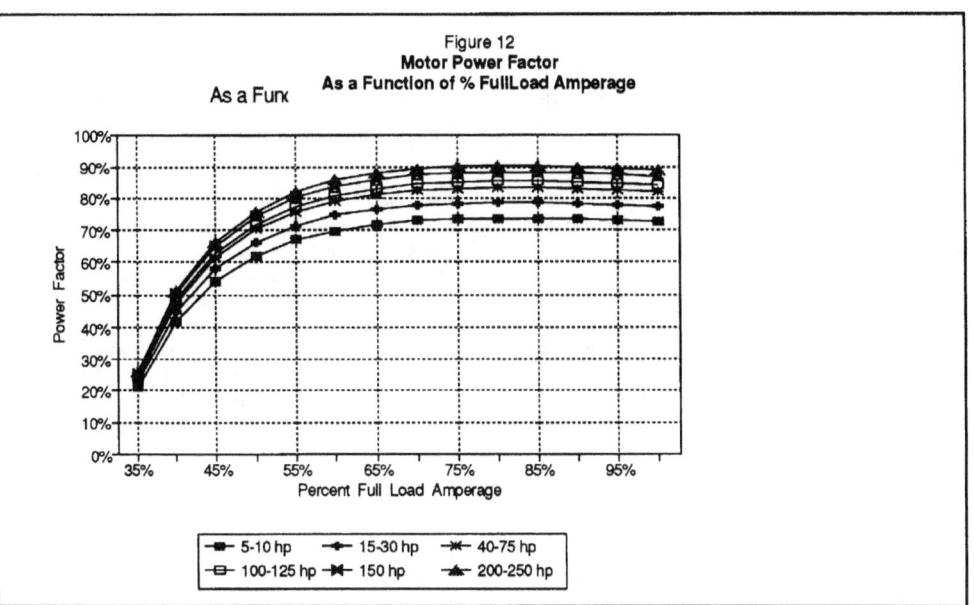

Figure 12
Motor Power Factor
As a Function of % FullLoad Amperage

Maximum efficiency does not usually* occur at full load As long as the motor is operating above 60 percent of rated load, the efficiency does not vary significantly.[22] Motor efficiencies typically improve down to about 75 percent of full-rated load, then, especially for smaller motors, rapidly begin to decline below 40 percent of full-load. It is almost always a good idea to downsize a motor that is less than 50 percent loaded.[23] Power factor declines sharply when the motor is operated below 75 percent of full-load amperage, especially in the smaller horsepower size ranges. Typical part-load motor efficiency and power factor characteristics are indicated in Figures 11 and 12.

The cost penalties associated with using an oversized motor can be substantial and include:[2]

- A higher motor purchase price
- Increased electrical supply equipment cost due to increased KVA and KVAR requirements
- Increased energy costs due to decreased part-load efficiency
- Power factor penalties

Replacing underloaded motors with smaller standard or energy-efficient motors improves efficiency.' Care must be taken, however, to fully understand the characteristics of the driven load before changing out existing motors.

For instance, with a variable load, such as a variable air volume HVAC system, the motor must be sized to operate under fully loaded conditions. Inlet vanes or other throttling devices must be set at "full open" so that efficiency and load factor measurements can be taken at maximum load.[23] Worn belts and pulleys can result in a reduced load being applied to the motor, giving the impression that it is underloaded. To eliminate this problem, worn belts or pulleys should be replaced before loan and efficiency tests are made.[23] Load types include:[19]

- Continuous-running steady loads
- Continuous-running with intermittent loading
- Variable-speed loads
- Cyclic loads

It is easiest to take measurements and properly size a motor driving a continuously-running steady load. Be sure to take torque characteristics into consideration for intermittent or cyclic loading patterns. Also, be sure to provide adequate fan circulation and cooling for motors coupled to adjustable-speed loads or variable speed drives. Overheating is a particular concern at either reduced or full-loads with the non-ideal voltage and current waveforms encountered with electronic variable-frequency drives.[24,25]

It is best to operate an induction motor at 65 to 100 percent of full-rated load You can save the most when the motor is properly matched with respect to the work that must be performed

The replacement of an operable standard-efficiency motor with a premium efficiency unit may be justified if the existing motor is operating inefficiently due to improper oversizing. In this instance, the cost effectiveness is bolstered due to the reduced cost of the smaller replacement motor and the substantial efficiency gain.

Electric load can be determined several ways. It can be directly measured with a power meter. A power factor meter (or Figure 12) and a clamp-on multimeter can be used in lieu of a power meter. Electric load can also be calculated from HP output. A stepwise procedure to determine HP output with only a tachometer is used on the data sheet in Figure 13. To determine electric load (i.e., input kW) the resulting HP output must be multiplied by 0.7457 and divided by the part load efficiency. Part load efficiency can be estimated from Figure 11, manufacturer, or MotorMaster data.

It is tempting to compute efficiency from HP ouput and measured electric load This is not recommended. Computed HP ouput can easily be in error by several percent, but efficiency requires accuracy within one or two percent. Efficiency varies significantly with temperature, running time, voltage, phase balance, and power quality. Attempts to obtain meaningful efficiency values from any sort of field testing have been discouraging.

For centrifugal loads, the replacement motor selected should be the next nameplate size above the motor output when operating under fully loaded conditions. It is recommended that voltage, amperage, kW draw, power factor, and slip be metered for a variety of motor operating conditions such that the maximum load point is known with confidence. **The slip technique should not be used for rewound motors or motors operating at other than their design voltage.**

Motors are selected based on startup, normal, or abnormal torque and load characteristics. This motor change out analysis approach is most useful for continuously operating motors under steady load conditions and for motors driving loads with low startup torque requirements, such as centrifugal fans and pumps where torque is a linear function of speed. The approach should not be used for motors driving conveyors or crushers-where oversizing may be required to account for high startup torque, transient loads, or abnormal operating conditions. **Most energy-efficient motors exhibit approximately the same locked rotor, breakdown, and rated load torque characteristics as their standard-efficiency counterparts.**

Figure 13
Oversized Motor Replacement Analysis

Motor Load and Operating Cost Analysis Form

Employee Name _____   Date _____

Company _____   Facility/Location _____

1. General Data

Application _____
What type of equipment the motor drives.

Energy rate _____ cents/kWh

Monthly Demand Rate _____

Annual Operating Hours _____

2. Motor Nameplate Data

Make/Model _____

Serial Number _____

Phase and Hz _____

Voltage Rating _____

Horsepower Rating _____

Enclosure Type _____

Synchronous Speed _____

Frame Size _____

NEMA Torque Class _____

Service Factor Rating _____

Insulation Class _____

Temperature Rise _____

FL Amperes Rating _____

FL RPM Rating _____

3. Measured Data

Input Volts _____
By Voltmeter

Input Amps _____
By Ampmeter

Input kW _____
By Powermeter if available, otherwise calculate below

Operating Speed _____
By Tachometer

4. Calculated Values

Full Load (FL) Slip _____
[Synchronous RPM - FL RPM Rating]

Operating Slip _____
[Synchronous RPM - Operating Speed]

Load Factor _____
[Operating Slip/FL Slip]

HP output _____
[Rated HP x Load Factor]

kVA input _____
[Input Volts x Input Amps x 0.001732]

Power factor _____
[(Average kW/kVA) x 100%, or based on Figure 2, using measured Input Amps/nameplate FL Amperage]

Input kW _____
[input Volts x input Amps x Power Factor x 0.001732]

Annual Operating Cost _____
[input kW x Annual Operating Hours x Energy Rate]

Annual Demand Cost (if any) _____
[Monthly Demand Rate x number of months motor operates during peak demand period]

Total Annual Operating Cost _____
[Annual Energy + Demand costs]

Chapter 6
Speed, Design Voltage, Enclosure, Part-load Efficiency, and Power Factor

Sensitivity of Efficiency Gains to Motor RPM

A motor's rotor must turn slower than the rotating magnetic field in the stator to induce an electrical current in the rotor conductor bars and thus produce torque. When the load on the motor increases, the rotor speed decreases. As the rotating magnetic field cuts the conductor bars at a higher rate, the current in the bars increases, which makes it possible for the motor to withstand the higher loading. Motors with slip greater than 5 percent are specified for high inertia and high torque applications.[24]

NEMA Design B motors deliver a starting torque that is 150 percent of full-load or rated torque and run with a slip of 3 to 5 percent at rated load.[24] Energy-efficient motors, however, are "stiffer" than equivalently sized standard motors and tend to operate at a slightly higher full-load speed. This characteristic is illustrated in Figure 14, which shows the full-load speed for 1,800 RPM standard and energy-efficient motors of various sizes. On the average, energy-efficient motors rotate only 5 to 10 RPM faster than standard models. The speed range for available motors, however, exceeds 40 to 60 RPM.

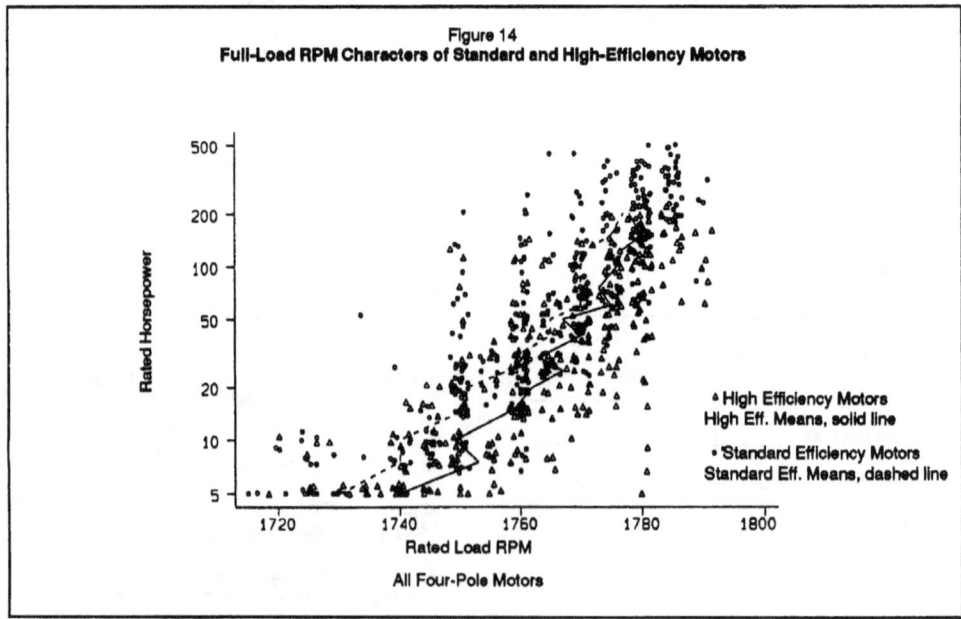

Figure 14
Full-Load RPM Characters of Standard and High-Efficiency Motors

▲ High Efficiency Motors
High Eff. Means, solid line

• Standard Efficiency Motors
Standard Eff. Means, dashed line

All Four-Pole Motors

Note: The solid lines indicate the mean or average speeds for all standard- and energy-efficient motors for each horsepower size, showing the higher typical RPM for energy-efficient motors.

For centrifugal loads, even a minor change in the motor's full-load speed translates into a significant change in the magnitude of the load and energy consumption. The "Fan" or "Affinity Laws," indicated in Table 15, show that the horsepower loading on a motor varies as the third power (cube) of its rotational speed. In contrast, the quantity of air delivered varies linearly with Speed.[26]

As summarized in Table 16, a relatively minor 20-RPM increase in a motor's rotational speed, from 1,740 to 1,760 RPM, results in a 3.5 percent increase in the load placed upon the motor by the rotating equipment. A 40-RPM speed increase will increase air or fluid flow by only 2.3 percent, but can boost energy consumption by 7 percent, far exceeding any efficiency advantages expected from purchase of a higher efficiency motor. Predicted energy savings will not materialize--in fact, energy consumption will substantially increase. This increase in energy consumption is especially troublesome when the additional air or liquid flow is not needed or useful.

Be aware of the sensitivity of load and energy requirements to rated motor speed. Replacing a standard motor with an energy-efficient motor in a centrifugal pump or fan application can result in increased energy consumption if the energy-efficient motor operates at a higher RPM. A standard-efficiency motor with a rated full-load speed of 1,750 RPM should be replaced with a high-efficiency unit of like speed in order to capture the full energy conservation benefits associated with a high-efficiency motor retrofit. Alternatively, you can use sheaves or trim pump impellers so equipment operates at its design conditions.

Table 15
Fan Laws/Affinity Laws

Law #1: $\dfrac{CFM_2}{CFM_1} = \dfrac{RPM_2}{RPM_1}$

Quantity (CFM) varies as fan speed (RPM)

Law #2: $\dfrac{P_2}{P_1} = \dfrac{(RPM_2)^2}{(RPM_1)^2}$

*Pressure (P) varies as the **square** of fan speed*

Law #3: $\dfrac{hp_2}{hp_1} = \dfrac{(RPM_2)^3}{(RPM_1)^3}$

Horsepower (hp) varies as the cube of fan speed

Table 16
Sensitivity of Load to Motor RPM

$\dfrac{(1,760)^3}{(1,740)^3} = 3.5 \text{ percent horsepower increase}$

$\dfrac{(1,780)^3}{(1,740)^3} = 7.0 \text{ percent horsepower increase}$

Operating Voltage Effects on Motor Performance

Generally, high-voltage motors have lower efficiencies than equivalent medium-voltage motors because increased winding insulation is required for the higher voltage machines. This increase in insulation results in a proportional decrease in available space for copper in the motor slot.[22] Consequently, I^2R losses increase.

Losses are also incurred when a motor designed to operate on a variety of voltage combinations (for example, 208 - 230/460 volts) is operated with a reduced voltage power supply. Under this condition, the motor will exhibit a lower full-load efficiency, run hotter slip more, produce less torque, and have a shorter life.[27] Efficiency can be improved by simply switching to a higher voltage transformer tap.

If operation at 208 Volts is required, an efficiency gain can be procured by installing an energy-efficient NEMA Design A motor. Efficiency, power factor, temperature rise, and slip are shown in Table 17 for typical open-drip proof 10 hp - 1800 RPM Design B and Design A motors operated at both 230 and 208 volts.[10,27]

Table 17
Performance Comparison for 10 hp NEMA Design B
Versus Design A Motors at 230 and 208 Volts

	Design B		Design A	
Volts	208	230	208	230
Efficiency, %	80.6	84.4	83.7	85.3
Power Factor, %	85.0	82.7	84.1	78.5
Temp. Rise, deg. C	91.0	72.0	73.0	66.0
Slip, %	5.9	4.1	4.6	3.5

Motor Speed and Enclosure Considerations

Energy-efficient motors are a worthwhile investment in all size, speed, and enclosure classifications. In general, higher speed motors and motors with open enclosures tend to have slightly higher efficiencies than low-speed or totally-enclosed fan-cooled units. In all cases, however, the energy-efficient motors offer significant efficiency improvements, and hence energy and dollar savings, when compared with the standard-efficiency models.

Typical motor efficiency gains are illustrated in Figures 15 through 17. Figure 15 shows the efficiency improvement expected from the selection of energy-efficient over standard-efficiency motors with varying nominal speeds. The efficiency gains are generally largest for the 3,600-RPM motors. Figure 16 indicates that the energy savings associated with 1,800-RPM energy-efficient over standard open motors slightly exceed those available from the high-efficiency over the standard totally enclosed model. Figure 17 indicates that energy-efficient motors provide even greater efficiency improvements when operating under part load conditions.

Efficiency Improvements at Part-Load Conditions

Energy-efficient motors perform better than their standard-efficiency counterparts at both full and partially loaded conditions. Typical efficiency gains for 5-, 20-, and 100-hp motors when operating at full; 3/4-; and 1/2-load are given in Figure 17. Efficiency improvements from use of a premium-efficiency motor actually increase slightly under half-loaded conditions. While the overall energy conservation benefits are less for partially versus fully-loaded motors, the percentage of savings remains relatively constant. To obtain full-, 3/4-, and 1/2-load efficiencies and power factor information, consult WSEO's Electric Motor Database.

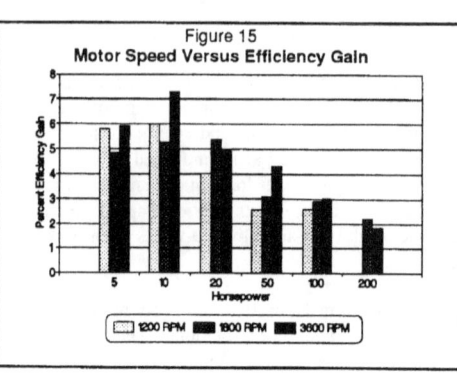

Figure 15
Motor Speed Versus Efficiency Gain

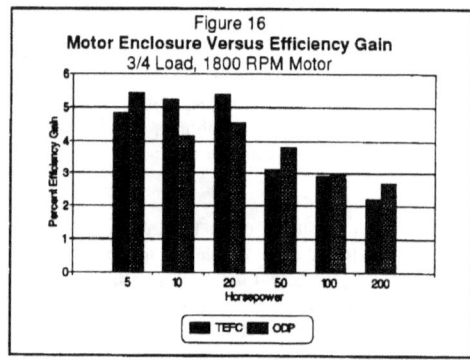

Figure 16
Motor Enclosure Versus Efficiency Gain
3/4 Load, 1800 RPM Motor

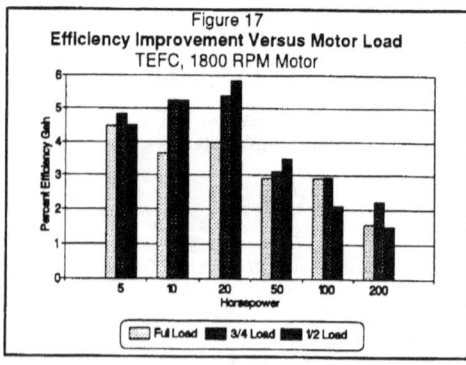

Figure 17
Efficiency Improvement Versus Motor Load
TEFC, 1800 RPM Motor

Power Factor Improvement

An induction motor requires both active and reactive power to operate. The active or true power, measured in kW, is consumed and produces work or heat. 'The reactive power, expressed in kVARs, is stored and discharged in the inductive or capacitive elements of the circuit., and establishes the magnetic field within the motor that causes it to rotate.[2,10] The total power or apparent power is the product of the total voltage and total current in an AC circuit and is expressed in KVA. The total power is also the vector sum of the active and reactive power components. Power factor is the ratio of the active to the total power, (Figure 18).

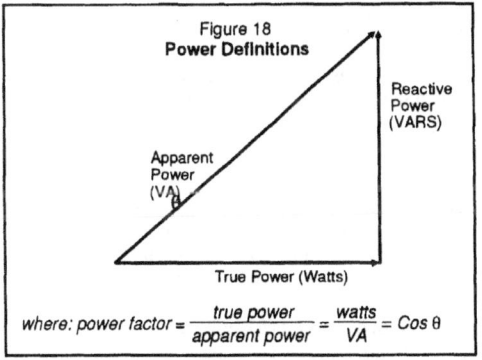

Figure 18
Power Definitions

where: power factor = $\frac{\text{true power}}{\text{apparent power}} = \frac{\text{watts}}{\text{VA}} = \cos\theta$

The electric utility must supply both active and reactive power loads. A low or "unsatisfactory" power factor is caused by the use of inductive (magnetic) devices and can indicate a possible low system electrical operating efficiency. Induction motors are generally the principal cause of low power factor because there are so many in use and they are not fully loaded.[28]

When motors operate near their rated load, the power factor is high, but for lightly loaded motors the power factor drops significantly. This effect is partially offset as the total current is less at reduced load. Thus, the lower power factor does not necessarily increase the peak KVA demand because of the reduction in load. Many utilities, however, levy a penalty or surcharge if a facility's power factor drops below 95 or 90 percent.

In addition to increased electrical billings, a low power factor may lower your plant's voltage, increase electrical distribution system line losses, and reduce the system's capacity to deliver electrical energy. While motor full- and part-load power factor characteristics are important, they are not as significant as nominal efficiency. When selecting a motor, conventional wisdom is to purchase efficiency and correct for power factor.[10]

Low power factors can be corrected by installing external capacitors at the main plant service or at individual pieces of equipment. Power factor can also be improved and the cost of external correction reduced by minimizing operation of idling or lightly loaded motors and by avoiding operation of equipment above its rated voltage.

Power factors can usually be improved through replacement of standard- with premium-efficiency motors. Power factors vary tremendously, however, based on motor design and load conditions. While some energy-efficient motor models offer power factor improvements of 2 to 5 percent, others have lower power factors than typical equivalent standard motors. Even high power factor motors are affected significantly by variations in load. A motor must be operated near its rated loading in order to realize the benefits of a high power factor design.

Chapter 7
Motor Operation Under Abnormal Conditions

Motors must be properly selected according to known service conditions. Usual service conditions, defined in NEMA Standards Publication MG1-1987, Motors and Generators, include:[29]

1. Exposure to an ambient temperature between 0°C and 40°C

2. Installation in areas or enclosures that do not seriously interfere with the ventilation of the machine

3. Operation within a tolerance of ± 10 percent of rated voltage

4. Operation from a sine wave voltage source (not to exceed 10 percent deviation factor)

5. Operation within a tolerance of ± 5 percent of rated frequency

6. Operation with a voltage unbalance of 1 percent or less

Operation under unusual service conditions may result in efficiency losses and the consumption of additional energy. Both standard and energy-efficient motors can have their efficiency and useful life reduced by a poorly maintained electrical system.[2] Monitoring voltage is important for maintaining high-efficiency operation and correcting potential problems before failures occur. Preventative maintenance personnel should periodically measure and log the voltage at a motor's terminals while the machine is fully loaded.

Over Voltage

As the voltage is increased, the magnetizing current increases by an exponential function. At some point, depending upon design of the motor, saturation of the core iron will increase and overheating will occur.[22] At about 10 to 15 percent over voltage both efficiency and power factor significantly decrease while the full-load slip decreases[2]. The starting current, starting torque, and breakdown torque all significantly increase with over voltage conditions.[15]

A voltage that is at the high end of tolerance limits frequently indicates that a transformer tap has been moved in the wrong direction. An overload relay will not recognize this over-voltage situation and, if the voltage is more than 10 percent high, the motor can overheat. Over voltage operation with VAR currents above acceptable limits for extended periods of time may accelerate deterioration of a motor's insulation.[24]

Under Voltage

If a motor is operated at reduced voltage, even within the allowable 10 percent limit, the motor will draw increased current to produce the torque requirements imposed by the load.[18] This causes an increase in both stator and rotor I^2R losses. Low voltages can also prevent the motor from developing an adequate starting torque. The effects on motor efficiency, power factor, RPM, and current from operating outside nominal design voltage are indicated in Figure 19.[14]

Reduced operating efficiency because of low voltages at the motor terminals is generally due to excessive voltage drops in the supply system.[2] If the motor is at the end of a long feeder, reconfiguration may be necessary. The system voltage can also be modified by:

- Adjusting the transformer tap settings
- Installing automatic tap-changing equipment if system loads vary considerably over the course of a day
- Installing power factor correction capacitors that raise the system voltage while correcting for power factor

Since motor efficiency and operating life are degraded by voltage variations, only motors with compatible voltage nameplate ratings should be specified for a system.

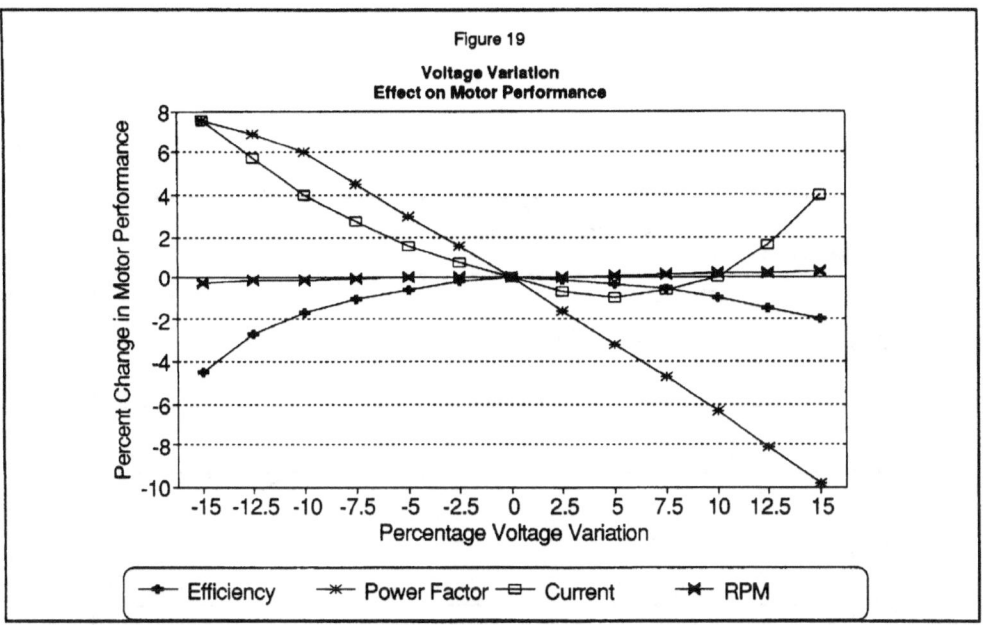

Figure 19
Voltage Variation Effect on Motor Performance

For example, three-phase motors are available with voltage ratings of 440, 460, 480, and 575 volts. The use of a motor designed for 460-volt service in a 480-volt system results in reduced efficiency, increased heating, and reduced motor life. A 440-volt motor would be even more seriously affected.[30]

Phase Voltage Imbalance

A voltage imbalance occurs when there are unequal voltages on the lines to a polyphase induction motor. This imbalance in phase voltages also causes the line currents to be out of balance. The unbalanced currents cause torque pulsations, vibrations, increased mechanical stress on the motor, and overheating of one and possibly two phase windings. This results in a dramatic increase in motor losses and heat generation, which both decrease the efficiency of the motor and shorten its life.[24]

Voltage imbalance is defined by NEMA as 100 times the maximum deviation of the line voltage from the average voltage on a three-phase system divided by the average voltage.[30] For example, if the measured line voltages are 462, 463, and 455 volts, the average is 460 volts. The voltage imbalance is:

$$\left(\frac{460 - 455}{460}\right) \times 100\% = 1.1\%.$$

A voltage unbalance of only 3.5 percent can increase motor losses by approximately 20 percent.[29] Imbalances over 5 percent indicate a serious problem. Imbalances over 1 percent require derating of the motor, and will void most manufacturers' warranties. Per NEMA MG1-14.35, a voltage imbalance of 2.5 percent would require a derate factor of 0.925 to be applied to the motor rating. Derating factors due to unbalanced voltage for integral horsepower motors are given in Figure 20[31]. The NEMA derating factors apply to all motors. There is no distinction between standard and energy-efficient motors when selecting a derate factor for operation under voltage unbalance conditions.

Common causes of voltage unbalance include:[2,24,30]

- Faulty operation of automatic power factor connection equipment
- Unbalanced or unstable utility supply
- Unbalanced transformer bank supplying a three-phase load that is too large for the bank

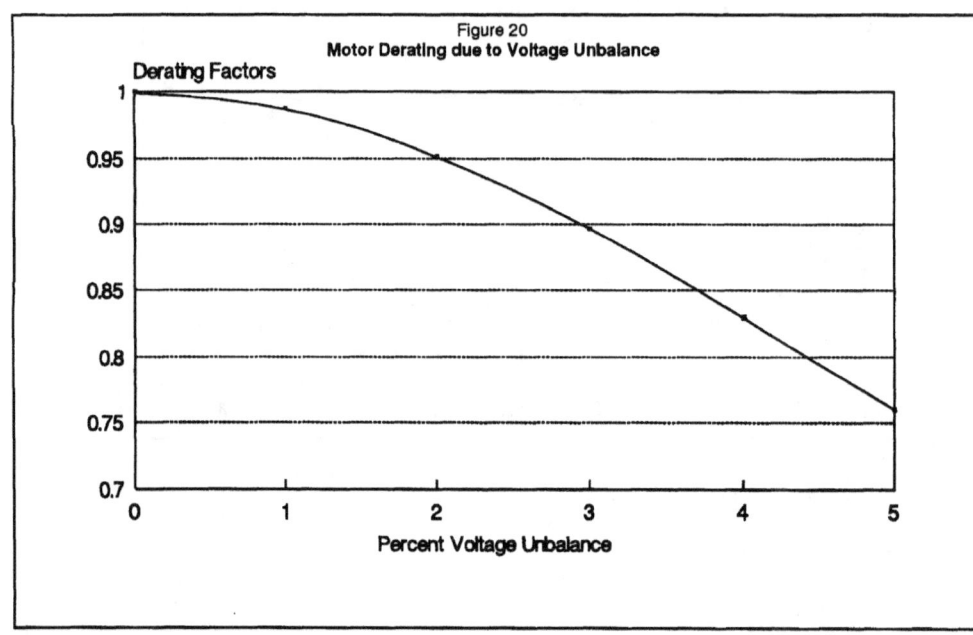

Figure 20
Motor Derating due to Voltage Unbalance

- Unevenly distributed single-phase loads on the same power system
- Unidentified single-phase to ground faults
- An open circuit on the distribution system primary

The following steps will ensure proper system balancing.[2]

- Check your electrical system single-line diagram to verify that single-phase loads are uniformly distributed
- Regularly monitor voltages on all phases to verify that a minimum variation exists.
- Install required ground fault indicators
- Perform annual thermographic inspections

Load Shedding

Energy and power savings can be obtained directly by shutting off idling motors to eliminate no-load losses.[1] This action also greatly improves the overall system power factor, which in turn improves system efficiency. Typical no-load or idling power factors are in the 10 to 20 percent range. Load shedding is most effective for slower speed (1,800 RPM and less) motors used in low-inertia applications.[10] While it is possible to save energy by de-energizing the motor and restarting it when required, excessive starting, especially without soft-starting capacity, can cause overheating and increased motor failures.

Consideration must be given to thermal starting capability and the life expectancy of both motor and starting equipment.[10] Motors 200 hp and below can only tolerate about 20 seconds of maximum acceleration time with each start Motors should not exceed more than 150 start seconds per day.[2] Starting limitations for motors over 200 hp should be obtained from the manufacturer. Maximum number of starts per hour and minimum off-time guidelines for 1800 RPM Design B motors of various sizes are given in Table 18.[29]

Table 18
Allowable Number of Starts and Minimum Time Between Starts (For 1800 RPM Design B Motors)

Motor Size, hp	Maximum Number of Starts per Hour[1]	Minimum Off Time (Seconds)
5	16.3	42
10	12.5	46
25	8.8	58
50	6.8	72
100	5.2	110

[1]This table is extracted from NEMA Standards Publications No. MG10 *Energy Management Guide for Selection and Use of Polyphase Motors.* NEMA has prepared a comprehensive load shedding table for 3600, 1800, and 1200 RPM motors in the 1- to 250-hp size range. NEMA also presents a methodology for minimizing winding stresses by adjusting the number of allowable starts per hour to account for load inertia.

Chapter 8
Motor Selection Considerations

Overall motor performance is related to the following parameters:[10]

- Acceleration capabilities
- Breakdown torque
- Efficiency
- Enclosure type
- Heating
- Inrush current
- Insulation class
- Power factor
- Service factor
- Sound level
- Speed
- Start torque

A good motor specification should define performance requirements and describe the environment within which the motor operates. As the purchaser, you should avoid writing design-based specifications that would require modification of standard components such as the frame, bearing, design, rotor design, or insulation class.[2]

Specification contents should include:
- Motor horsepower and service factors
- Temperature rise and insulation class
- Maximum starting current
- Minimum stall time
- Power factor range
- Efficiency requirement and test standard to be used
- Load inertia and expected number of starts

Environmental information should include:
- Abrasive or non-abrasive
- Altitude
- Ambient temperature
- Hazardous or non-hazardous
- Humidity level

You should specify special equipment requirements such as thermal protection, space heaters (to prevent moisture condensation), and whether standard or non-standard conduit boxes are required.

Motor Enclosures

Many types of motor enclosures are available, including:[16]

Open. An enclosure with ventilating openings that permit passage of external cooling air over and around the motor windings. This design is now seldom used.

Open Drip-Proof (ODP). An open motor in which ventilation openings prevent liquid or solids from entering the machine at any angle less than 15 degrees from the vertical.

Guarded. An open motor in which all ventilating openings are limited to specified size and shape. This protects fingers or rods from accidental contact with rotating or electrical parts.

Splash-Proof. An open motor in which ventilation openings prevent liquid or solids from entering the machine at any angle less than 100 degrees from the vertical.

Totally-Enclosed. A motor enclosed to prevent the free exchange of air between the inside and outside of the case, but not airtight.

Totally-Enclosed Nonventilated (TENV). A totally-enclosed motor that is not equipped for cooling by means external to the enclosed parts.

Totally-Enclosed Fan-Cooled (TEFC). A totally-enclosed motor with a fan to blow cooling air across the external frame. They are commonly used in dusty, dirty, and corrosive atmospheres.

Encapsulated. An open motor in which the windings are covered with a heavy coating of material to provide protection from moisture, dirt, and abrasion.

Explosion-Proof. A totally-enclosed motor designed and built to withstand an explosion of gas or vapor within it, and to prevent ignition of gas or vapor surrounding the machine by sparks, flashes, or explosions that may occur within the machine casing.

Motor Insulation Systems

The ultimate cause of motor failure is frequently internal heat production and increased operating temperatures due to high currents or contamination. An insulation system is comprised of insulating materials for conductors and the structural parts of a motor.[16,32] Since motor failure often occurs due to oxidation and thermal degradation of insulating materials, motors that run hotter tend to have shorter operating lives. The relationship between operating temperature and motor insulation life is shown in Figure 21.[27] A typical rule of thumb is that the service life expectancy of a motor is reduced by one-half for each 10°C increase in operating temperature.

Grease life also varies with temperature. As the bearing temperature increases, a motor must be regreased more frequently to prevent premature bearing failures.[19]

All insulation systems are not the same. NEMA has established standards for insulation design, temperature rating, and motor thermal capacity.[29] Four classes of insulation have been designated, each with an allowable operating temperature. These insulation systems, designated classes A, B, F, and H, vary with respect to design and selection of material and bonding agent thermal range. A Class A insulation system is one which is shown by experience or test to have a suitable operating life when operated at 105°C. A Class B system shows acceptable thermal endurance when operated at 130°C; a Class F insulation system can be operated at 155°C, while a Class H system can be operated at a limiting temperature of 180° C.[16] Class B and F systems are most commonly used.

Service Factor

Motors are designed with an allowable increase in temperature above ambient during operation. This is referred to as temperature rise. The maximum allowable temperature rise during operation for a motor varies with respect to insulation class and the motor's service factor. The service factor is essentially a safety margin and refers to the motor's ability to continuously deliver horsepower beyond its nameplate rating under specified conditions. Most motors are rated with a 1.0 or 1.15 service factor. A 10-hp motor operating under rated conditions with a 1.15 service factor should be able to continuously deliver 11.5 horsepower without exceeding the NEMA allowable temperature rise for its insulation system.[29] NEMA allows an ambient temperature of 40°C (104°F) when specifying "usual service conditions."

If the ambient temperature exceeds 40° C, or at elevations above 3,300 feet, the motor service factor must be reduced or a higher horsepower motor used As the oversized motor will be underloaded, the operating temperature rise is less and overheating will be reduced.[16]

NEMA temperature standards for motors with Class B and F insulation and a 1.0 or 1.15 service factor are given in Table 19.[2] Note that a motor equipped with Class F insulation, but operating within class B temperature limitations, is operating far below its maximum operating limitations. It is thus running "cooler" relative to its thermal capability.[29] Premium- or energy-efficient motors are typically equipped with Class F insulation and rated with a 1.15 service factor.

Table 19
Temperature Limitations for Insulation Classes

Service Factor	Insulation Temperature	Class B	Class F
1.0/1.15	Ambient Temperature	40°C/104°F	40°C/104°F
1.0	Allowable Temperature Rise	80°C/176°F	105°C/221°F
1.0	Operating Temperature Limitation	120°C/248°F	145°C/293°F
1.15	Allowable Temperature Rise	90°c/194°F	115°C/239°F
1.15	Operating Temperature Limitation	130°C/266°F	155°C/311°F

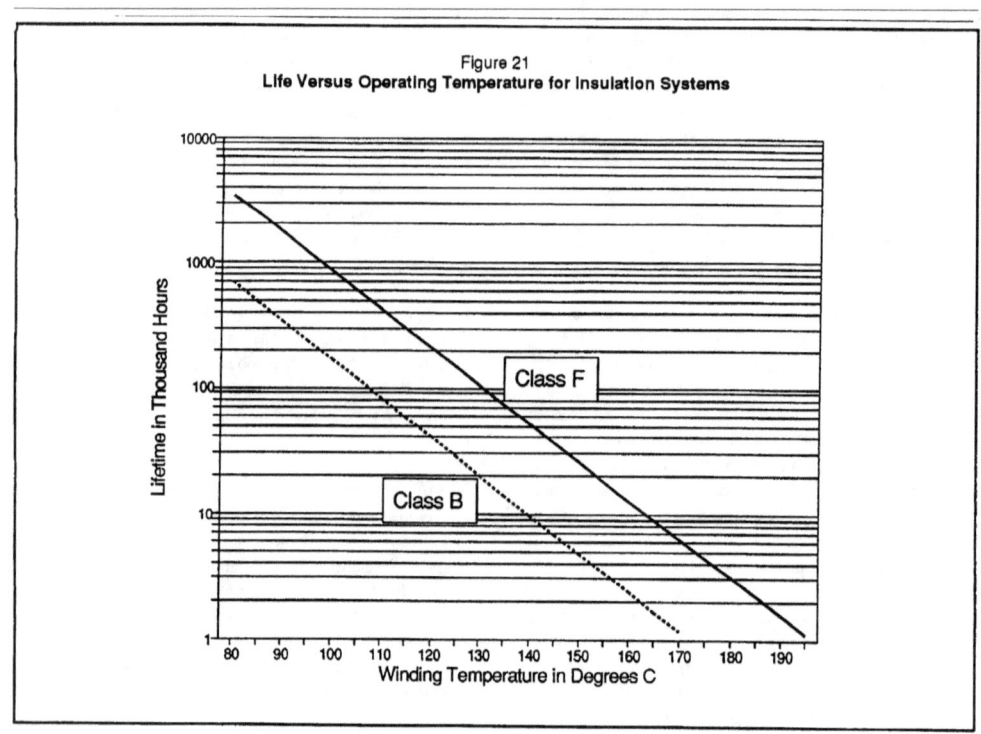

Figure 21
Life Versus Operating Temperature for Insulation Systems

Motor Speed, Slip, and Torque Characteristics

When selecting the proper motor speed, consider the original equipment cost and the requirements of the driven system. Generally, large high-speed standard or energy-efficient motors have improved efficiency and power factor characteristics.

Load, torque, and horsepower requirements determine the type and size of motor required for a particular application. Torque is a measure of the rotational force that a motor can produce. As the physical size of a motor is proportional to its torque capability, high-torque motors are larger and cost more.[2]

Induction motors are standardized according to their torque characteristics (Design A, B, C, and D).[2] Torque is in turn characterized by starting or locked-rotor torque, which is the minimum torque produced by the motor at rated voltage and frequency at all angular positions of the rotor; pull-up torque, which is the minimum torque developed by the motor during acceleration; and breakdown torque, which is the maximum torque that the motor can supply before stalling. A representative speed-torque curve for a Design B induction motor is shown in Figure 22.[16]

The motor design selected must have adequate torque capability to start a load and accelerate it to full speed NEMA Design B motors can be used with constant speed centrifugal fans, pumps and blowers, unloaded compressors, some conveyors, and cutting machine tools.[16] Most induction motors are Design B, with Design A being the second most common.[2] While NEMA limits for locked rotor torque for Design A and B motors are the same, some manufacturers design their motors to different criteria. Frequently, Design A motors have higher starting current and start-up torque characteristics. Speed-torque characteristics for polyphase motors are given in Table 20.[29]

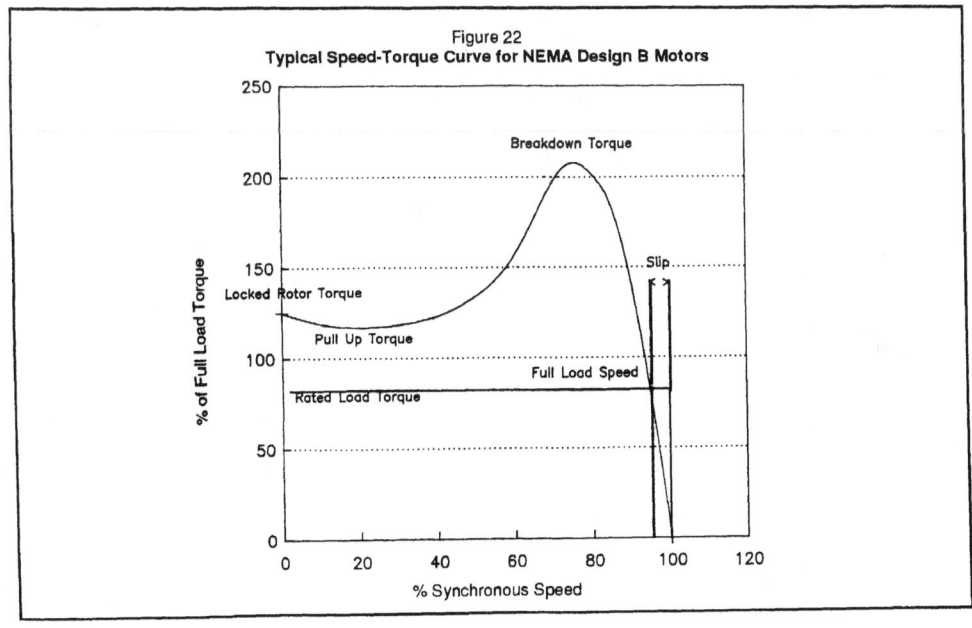

Figure 22
Typical Speed-Torque Curve for NEMA Design B Motors

Table 20
NEMA Torque Characteristics for Medium Polyphase Induction Motors[1]

NEMA Design	Starting Current (% Rated Load Current)	Locked Rotor Torque (% Rated Load Torque)	Breakdown Torque (% Rated Load Torque)	Percent slip
B[2]	600-700	70 - 275	175 - 300	1 - 5
C	600-700	200 - 250	190 - 225	5
D	600-700	275	275	5 - 8

[1] NEMA Standards Publication for Motors and Generators MG1-1978 classifies motors as medium or large. 3600 or 1800 RPM motors rated up to 500 hp are defined as medium motors. The rating declines to 350 and 250 hp for 1200 and 900 RPM motors, respectively.

[2] Design A motor have characteristics similar to those for Design B motors except that starting currents are higher.

Motor service lifetimes can be extensive, typically exceeding 10 years, when the unit is properly matched to its driven load and operated under design power supply conditions. Historically, the single largest cause of motor failure has been overloading due to improper matching of motors to the load or placing motors into operation under conditions of voltage imbalance. Causes of failure include:[33]

Overload (overheating)	25%
Contamination	
Moisture	17%
Oil and grease	20%
Chemical	1%
Chips and dust	5%
Single Phasing	10%
Bearing Failure	12%
Normal Insulation	
Deterioration	5%
Other	5%

Chapter 9
Additional Benefits of Energy-Efficient Motors

Energy-efficient motors are longer than standard-efficiency motors as the rotor and stator cores are lengthened to reduce losses associated with the magnetic flux density. However, they are mounted in the same frame as corresponding standard-efficiency T-frame motors. They fully conform with NEMA inrush current, starting, and breakdown torque standards. Conventional NEMA controls and protection can be applied.[11]

Energy-efficient motors typically operate cooler than their standard efficiency counterpart? Lower operating temperatures translate into increased motor, insulation, and bearing life. The result is fewer winding failures, increased bearing life, longer periods between scheduled maintenance actions, and fewer forced outages.

Accelerated life testing, by subjecting the motor to repeated stalls and other abuse, indicates that energy-efficient motors should have a longer life expectancy than standard-efficiency designs.[20] Besides this increased capacity to withstand stalling and overloads, energy-efficient motors should run quieter and operate with lower no-load losses.

Besides reducing operating costs and extending winding and bearing service lives, additional benefits typically associated with using energy-efficient motors include:[11,30]

- An extended warranty
- Extended lubrication cycles due to cooler operation
- Better tolerance to thermal stresses resulting from stalls or frequent starting
- The ability to operate in higher ambient temperatures
- Increased ability to handle overload conditions due to cooler operation and a 1.15 service factor
- Fewer failures under conditions of impaired ventilation
- More resistance to abnormal operating conditions, such as under and over voltage or phase unbalance
- More tolerance to poorer voltage and current waveshapes
- A slightly higher power factor in the 100 hp and lower size range, which reduces distribution system losses and utility power factor penalty changes

These benefits however, depend on many factors. Based on manufacturer design practices, energy-efficient motors may have higher or lower power factors than their standard-efficiency counterparts. Both energy-efficient and standard motors should be derated the same amount under conditions of voltage unbalance. Generally, the perception exists that standard and energy-efficient motors operate at different temperatures and there is more temperature margin available in the energy-efficient motor before reaching NEMA operating temperature limits.

Chapter 10
Starting Your Motor Improvement Program

Begin your motor improvement program by screening to select the best candidates for immediate retrofit or future replacement with energy-efficient units. Complete an Electric Motor Data Sheet for each motor used in excess of 2,000 hours per year. Examine rewind history, motor age, and application.[20] A recording wattmeter may be useful for analyzing varying loads over a representative period of time.

Check with the local utility and the Bonneville Power Administration regarding the availability of financial incentives such as high-efficiency electric motor rebates or commercial/industrial sector energy conservation programs, technical assistance, billing credit offers, or competitive bidding solicitations.

If financial incentives are available, it may be it cost-effective to complete a "group" conversion of eligible motors rather than wait for operable standard-efficiency motors to fail. A disadvantage of waiting until failure to replace motors is that energy-efficient units might not be readily available. Also, once a motor fails, it is no longer possible to check whether the motor is properly matched to the load.[23] Too often, immediate replacement needs outweigh energy management objectives and a failed standard-efficiency motor is replaced by a standard efficiency spare. Immediate replacement locks in future energy savings at today's capital cost.[20]

Replacement of an operable standard-efficiency motor may also be wise from a preventative maintenance standpoint. Some would argue that "...the most efficient motor is one that runs consistently...thus reducing downtime to an absolute minimum."[32] Replacement of aged standard-efficiency motors with new, more reliable energy-efficient units should provide secondary economic benefits through prevention of unexpected failures and increased productivity.

Downtime costs can be substantial. Estimates of typical downtime costs for various industries are summarized in Table 21.[32]

Table 21
Estimated Downtime Costs for Selected Industries

Industry	Cost
Forest Industries	$7,000/hour
Food Processing	$30,000/hour
Petroleum & Chemical	$87,000/hour
Metal Casting	$100,00/hour
Automotive	$200,000/hour

To obtain specific motor performance information or assistance in evaluating your motor selection alternatives, contact the Electric Ideas Clearinghouse at 1-800-872-3568 or 206-586-8588 outside of BPA's service territory. The Clearinghouse is funded by BPA and operates through the Washington State Energy Office.

Chapter 11
Energy-Efficient Motors:
Twenty Questions and Answers

To help you assess the viability of energy-efficient motors for your operation, here are answers to typical technical, and financial questions. These questions are adapted from B.C. Hydro's Power Smart Publication entitled *High-Efficiency Motors*.

1. What is an energy-efficient motor?

An energy-efficient motor produces the same shaft output power (hp), but uses less electrical input power (kW) than a standard-efficiency motor. Energy-efficient motors must have nominal full-load efficiencies that exceed the minimum NEMA standards given in Table 2.

Many motor manufacturers produce models that significantly exceed the NEMA standard. These may be called "high" or "premium" efficiency motors.

2. How is an energy-efficient motor different than a standard motor?

Energy-efficient motors are manufactured using the same frame as a standard T-frame motor, but have:

- Higher quality and thinner steel laminations in the stator
- More copper in the windings
- Smaller air gap between the rotor and stator
- Reduced fan losses
- Closer machining tolerances

3. Are all new motors energy-efficient motors?

No, you generally have to ask for them.

4. Where can I buy a energy-efficient motor?

Energy-efficient motors can be purchased directly from most motor distributors. They can also be specified in any equipment package you may be ordering.

5. Are energy-efficient single-phase motors available?

Although energy-efficient single-phase motors are not common, a few manufacturers are beginning to produce them.

6. Do energy-efficient motors require more maintenance?

No. Energy-efficient motors have the same maintenance requirements as standard motors and are often more reliable.

7. What hp, speed, and voltage ranges are available?

Energy-efficient motors are available for most motor sizes 1 hp and up at speeds of 3600, 1800, 1200, and 900 RPM and three-phase voltages of 208, 230, 460, 575, and higher.

8. Can a energy-efficient motor replace my present U- or T-frame motor?

Yes. Since T-frame energy-efficient motors generally use the same frame casting as a standard motor, standard T-frame to high-efficiency T-frame should be a straight replacement. An adapter or transition base is required for a U-frame to energy-efficient T-frame replacement. In addition, some manufacturers now make energy-efficient U-frame motors. Talk to motor dealers for specifics.

9. Should I rewind my standard efficiency motor or purchase an energy-efficient motor?

An energy-efficiency motor will result in lower energy costs when compared with a rewound motor. Its cost effectiveness will depend on the hours operated, motor efficiencies, utility rates, and the difference in cost between the rewind and the energy-efficient motor. (Guidelines for energy-efficient motor purchase versus rewinds are given in Table 14.)

10. Can a standard motor be rewound as an energy-efficient motor?

It is possible for a standard motor to have what is commonly called a "high-efficiency rewind." This rewind procedure can slightly increase the efficiency of a standard motor above its initial level. However, the efficiency would still be lower than that of a new energy-efficient motor because of its unique physical characteristics. Energy-efficient motors can also be rewound.

11. What is the efficiency of a energy-efficient motor at different load points?

The efficiency of any motor varies with such factors as size, speed, and loading. As indicated in Figure 17, energy-efficient motors offer performance improvements over standard efficiency motors under full, partial, and unloaded conditions.

12. Do energy-efficient motors maintain the same percentage edge over standard motors when the load range drops from full load?

Yes. Most manufacturers are designing their energy-efficient motors to provide peak efficiency at 75 percent to 100 percent load. As shown in Figures 11 and 17, efficiency stays fairly constant from full down to 50 percent load, but the power factor drops significantly.

13. How reliable are energy-efficient motors?

Energy efficient motors are as reliable as regular or standard-efficiency motors. In some cases, they have a longer life because of lower motor operating temperatures.

14. What is the power factor of an energy-efficient motor?

Power factors vary tremendously depending on motor loading and manufacturer. While some energy-efficient motor models offer power factor improvements of 2 to 5 percent, others have lower power factors than their standard motor counterparts. On the average, a power factor improvement of less than 1 percent is expected.

15. I have heard different types of efficiencies quoted. What are they?

The following motor efficiency definitions are used: Quoted, Nominal, Average, Expected, Calculated, Minimum, Guaranteed, and Apparent. The most commonly used are Nominal and Minimum, defined as:

- Nominal Efficiency is the average measured efficiency of a large number of motors of the same design.
- Minimum Efficiency is the value appearing at the end of the bell curve plotted from the measurement of a large number of motors.

16. What are NEMA MG1, IEC 34.2, and JEC 37?

These are motor efficiency test or product standards:

- NEMA MG1, based upon the IEEE Standard 112 Method B motor efficiency testing methodology, is the most commonly used North American standard.
- CSA C390-M1985 is a Canadian-developed standard that is more rigorous than other test standards.
- IEC 34.2 is the European motor test standard.
- JEC 37 is the Japanese motor test standard.

17. Can I compare motor efficiencies using nameplate data?

Per NEMA MG1 - 12.54.2 the efficiency of Design A and Design B motors in the 1- to 125-hp range for frames in accordance with MG 13 shall be marked on the motor nameplate. As nameplate nominal efficiencies are rounded values, you should always obtain efficiency values from the motor manufacturer.

18. Is the service factor any different from that of a standard motor?

The service factor for many energy-efficient motors is at least 1.15 and can be as high as 1.30 to 1.40.

19. How much do energy-efficient motors cost?

Generally, they cost 15 to 30 percent more than standard motors, depending on the specific motor, manufacturer, and market competition It may, however, be possible to negotiate a lower price premium when purchasing a large quantity of energy-efficient motors.

20. What is the payback period for selecting a energy-efficient versus a standard efficiency motor?

The payback period varies according to the purchase scenario under consideration, cost difference, hours of operation, electrical rates, motor loading, and difference in motor efficiencies. For new purchase decisions, the simple payback on the incremental cost of a continuously operated energy-efficient motor can be recovered through energy savings in less than 2 years.

Chapter 12
References

1. South Carolina Governor's Division of Energy, Agriculture and Natural Resources. Energy *Conservation Manual.*

2. B.C. Hydro. "High-Efficiency Motors." Power Smart Brochure.

3. Northwest Power Planning Council. 1989. *1989 Supplement to the Northwest Conservation and Electric Power Plan, Volume II.*

4. Ciliano, Robert, et al. *Industrial Sector Conservation Supply Curve Data Base - Executive Summary.* Synergic Resources Corp. October 1988.

5. Reliance Electric. *A-C Drives: A Key to Reducing Operating Costs and Increasing Productivity.*

6. Seton, Johnson & Odell. *Industrial Motor Drive Study.* June 1983.

7. Bonnett, Austin H. of U.S. Electrical Motors. Letter to Gilbert McCoy of WSEO. February 7, 1990.

8. National Electrical Manufacturers Association. Publication MG1, News Release. March 1989.

9. Keinz, John R. and R.L. Hotulton. "NEMA/Nominal Efficiency: What is it and why?" *IEEE Conference Record CH1459-5,* Paper No. PCI-80-8 1980.

10. Bonnett, Austin H. "Understanding Efficiency and Power Factor in Squirrel Cage Induction Motors." U.S. Electrical Motors, A Presentation to the Washington State Energy Office. April 1990.

11. Montgomery, David C. *How to Specify and Evaluate Energy-Efficient Motors.* General Electric Company.

12. Bonnet, Austin H. *Understanding Power Factor in Squirrel Cage Induction Motors.* U.S. Electrical Motors.

13. Lovins, Amory B., et al. "State of the Art: Drivepower." Rocky Mountain Institute, Snowmass, Colorado. April 1989.

14. From K.K. Lobodovsky, Pacific Gas & Electric Company, San Francisco, California

15. Lobodovsky, K.K. *Electric Motors: Premium versus Standard.* Pacific Gas & Electric Company San Francisco, California..

16. The Lincoln Electric Company *Fundamentals of Polyphase Electric Motors.* Cleveland, Ohio. April 1987.

17. Stebbins, W.L., Electrical Senior Staff Engineer Hoechst Celanese Textile Fibers, Rock Hill, South Carolina Presented in an Energy Management short course entitled "Energy Auditing and Analysis of Industrial Commerical Facilities. "University of Wisconsin - Madison. 1990.

18. McCoy, Gilbert A. and Kim Lyons. *Local Government Energy Management: High Efficiency Electric Motor Applications.* Washington State Energy Office, WAOENG-83-49. December 1983.

19. Montgomery D.C. *Avoiding Motor Efficiency Degradation.* Presented at the 7th World Energy Engineering Congress, Atlanta, Georgia. November, 1984.

20. McGovern, William U. "High-Efficiency Motors for Upgrading Plant Performance." *Electric Forum,* Vol. 10, No. 2 1984.

21. Lobodovsky, Konstantin of Pacific Gas & Electric Company and Ramesh Ganeriwal and Anil Gupta of the California Energy Commission. *Field Measurements and Determination of Electric Motor Efficiency.* Presented at the Sixth World Energy Engineering Congress, Atlanta, Georgia December, 1983.

22. Bonnett, Austin H. *Understanding Efficiency in Squirrel Cage Induction Motors.* U.S. Electrical Motors.

23. Wilke K. and T. Ikuenobe. *Guidelines for Implementing an Energy-Efficient Motor Retrofit Program.* Presented at the 10th World Energy Engineering Congress, Atlanta, Georgia. October 1987.

24. The Electrification Council. *Motors and Motor Controls,* Third Edition. 1986, Reprint, 1989.

25. Ebasco Services, Inc. *Adjustable Speed Drive Applications Guidebook.* Prepared for the Bonneville Power Administration January 1990.

26. Lobodovsky, K.K. "Fan Applications: Fan Types and Fan Laws." Pacific Gas & Electric Technical Services Application Note No. 23-67-84.

27. Bonnett, Austin H. and L.R. Laub. *A Case For Design 'A' Induction Motors.* U.S. Electrical Motors.

28. Lobodovsky, K.K. "Power Factor Correction Capacitors - What - Why - Where - How." Pacific Gas & Electric Technical Services Application Note No. 43-46-83.

29. NEMA Standards Publication No. MG10. *Energy Management Guide for Selection and Use of Polyphase Motors.* National Electrical Manufacturers Association, Washington, D.C. 1989.

30. Jones, Randy. Federal Department of Energy. Transfer of Energy Efficiency Drivepower Project. Draft working paper. March 22, 1991.

31. Nadel, Steven. et.al. *Energy-Efficient Motor Systems: A Handbook on Technology, Programs, and Policy Opportunities. American Council for an Energy-Efficient Economy.* 1991.

32. Revelt, Jean. *Evaluating Electric Motors.* Lincoln Electric Company, Cleveland. Ohio.

33. Andreas, J.C. *Energy-Efficient Electric Motors: Selection and Application.* Marcel Decker Inc. 1982.

Appendix A

Motor Load and Operating Cost Analysis Form

Employee Name _____   Date _____

Company _____   Facility/Location _____

1. General Data

Application _____
What type of epuipment the motor drives.

Energy rate _____ cents/kWh

Monthly Demand Rate _____

Annual Operating Hours _____

2. Motor Nameplate Data

Make/Model _____

Serial Number _____

Phase and Hz _____

Voltage Rating _____

Horsepower Rating _____

Enclosure Type _____

Synchronous Speed _____

Frame Size _____

NEMA Torque Class _____

Service Factor Rating _____

Insulation Class _____

Temperature Rise _____

FL Amperes Rating _____

FL RPM Rating _____

3. Measured Data

Input Volts _____
By Voltmeter

Input Amps _____
By Ampmeter

Input kW _____
By Powermeter if available, otherwise calculate below

Operating Speed _____
By Tachometer

4. Calculated Values

Full Load (FL) Slip _____
[Synchronous RPM - FL RPM Rating]

Operating Slip _____
[Synchronous RPM - Operating Speed]

Load Factor _____
[Operating Slip/FL Slip]

HP output _____
[Rated HP x Load Factor]

kVA input _____
[Input Volts x Input Amps x 0.001732]

Power factor _____
[(Average kW/kVA) x 100%, or based on Figure 2, using measured Input Amps/nameplate FL Amperage]

Input kW _____
[input Volts x input Amps x Power Factor x 0.001732]

Annual Operating Cost _____
[input kW x Annual Operating Hours x Energy Rate]

Annual Demand Cost (if any) _____
[Monthly Demand Rate x number of months motor operates during peak demand period]

Total Annual Operating Cost _____
[Annual Energy + Demand costs]

Appendix B

Motor Manufacturer Address List

A.O. Smith
531 North Fourth
Tippcity, OH 45371
(513) 667-6800
FAX (513) 667-5873

Baldor
PO Box 240
Fort Smith, AZ 72902
(501) 646-4711
FAX: (501) 648-5792

Brook Crompton Inc.
3186 Kennicott Avenue
Arlington Heights, IL 60004
(708) 253-5577
FAX: (708) 253-9880

Dayton/Grainger
5959 W. Howard
Chicago, IL 60648
(800) 323-0620
FAX: (800) 722-3291

General Electric
P.O. Box 2222
Fort Wayne, IN 46801
(219) 428-2000
FAX: (219) 428-2731

Leeson
2100 Washington Ave.
Grafton, WI 53024
(414) 377-8810
FAX: (414) 377-9025

Lincoln/Delco
22801 St Clair Ave.
Cleveland, OH 44117
(216) 481-8100
FAX: (216) 486-1751

MagneTek / Century / Louis Allis
1881 Pine St.
St Louis, MO 63103
(800) 325-7344
FAX: (800) 468-2045

Marathon
P.O. Box 8003
Wausau, WI 54402
(715) 675-3311
FAX: (715) 675-9413

Reliance
24701 Euclid Ave.
Cleveland, OH 44117
(800) 245-4501
FAX: (800) 266-7536

Siemens
4620 Forest Ave.
Norwood, OH 45212
(513) 841-3100
FAX: (513) 841-3407

Sterling
16752 Armstrong Ave.
Irvine, CA 92714
(800) 654-6220
FAX: (714) 474-0543

Teco American
6877 Wynnwood
Houston, TX 77008
(713) 864-5980
FAX: (713) 865-7627

Toshiba
13131 W. Little York Rd.
Houston, TX 77041
(713) 466-0277
FAX: (713) 466-8773

US Motors / Emerson / Leroy Somer
8000 W. Florissant Ave.
St. Louis, MO 63136
(314) 553-2000
FAX: (314) 553-3527

Westinghouse Motors
IH-35 Westinghouse Road
P.O. Box 277
Rockround, TX 78680-0277
(512) 255-4141
FAX (512) 244-5512

For further information, contact one of the following U.S. Department of Energy support offices:

Atlanta	(404) 347-2837
Chicago	(708) 252-2208
Denver	(303) 231-5750
Kansas City	(816) 426-5533
Philadelphia	(215) 597-3890
San Francisco	(510) 273-4461
Boston	(617) 565-9700
Dallas	(214) 767-7245
New York	(212) 264-1021
Seattle	(206) 553-1004

Produced by the Washington State Energy Office for the Bonneville Power Administration, reprinted with permission for the U.S. Department of Energy.

www.ingramcontent.com/pod-product-compliance
Lightning Source LLC
Chambersburg PA
CBHW071811170526
45167CB00003B/1260